Lecture Notes in Computational Science and Engineering

27

Editors

T. J. Barth, Moffett Field, CA
M. Griebel, Bonn
D. E. Keyes, Norfolk
R. M. Nieminen, Espoo
D. Roose, Leuven
T. Schlick, New York

T0223857

Springer
Berlin
Heidelberg
New York
Hong Kong
London
Milan
Paris
Tokyo

Siegfried Müller

Adaptive Multiscale Schemes for Conservation Laws

With 58 Figures

Springer

Siegfried Müller
Institut für Geometrie
und Praktische Mathematik
RWTH Aachen
Templergraben 55
52056 Aachen, Germany
e-mail: mueller@igpm.rwth-aachen.de

Cataloging-in-Publication Data applied for

Bibliographic information published by Die Deutsche Bibliothek

Die Deutsche Bibliothek lists this publication in the Deutsche Nationalbibliografie; detailed bibliographic data is available in the Internet at <http://dnb.ddb.de>.

Mathematics Subject Classification (2000):
65M12, 65M55, 42C15, 47A20, 76Axx, 35L65

ISSN 1439-7358
ISBN 3-540-44325-8 Springer-Verlag Berlin Heidelberg New York

Springer-Verlag Berlin Heidelberg New York
a member of BertelsmannSpringer Science + Business Media GmbH
http://www.springer.de

Printed in Germany

Cover Design: Friedhelm Steinen-Broo, Estudio Calamar, Spain
Cover production: *design & production*
Typeset by the author using a Springer TEX macro package

Printed on acid-free paper SPIN: 10885274 46/3142/LK - 5 4 3 2 1 0

To my parents

Preface

During the last decade enormous progress has been achieved in the field of computational fluid dynamics. This became possible by the development of robust and high–order accurate numerical algorithms as well as the construction of enhanced computer hardware, e.g., parallel and vector architectures, workstation clusters. All these improvements allow the numerical simulation of *real world* problems arising for instance in automotive and aviation industry. Nowadays numerical simulations may be considered as an indispensable tool in the design of engineering devices complementing or avoiding expensive experiments. In order to obtain qualitatively as well as quantitatively *reliable* results the complexity of the applications continuously increases due to the demand of resolving more details of the real world configuration as well as taking better physical models into account, e.g., turbulence, real gas or aeroelasticity. Although the speed and memory of computer hardware are currently doubled approximately every 18 months according to Moore's law, this will not be sufficient to cope with the increasing complexity required by *uniform* discretizations.

The future task will be to *optimize* the utilization of the available resources. Therefore new numerical algorithms have to be developed with a computational complexity that can be termed nearly optimal in the sense that storage and computational expense remain proportional to the "inherent complexity" (a term that will be made clearer later) problem. This leads to adaptive concepts which correspond in a natural way to *unstructured* grids. The conclusion is justified by results of approximation theory which clearly indicate that nonlinear approximations, e.g., the positions of the discretization points are not a priori fixed, are more efficient than linear approximations, e.g., uniform discretizations. For details on nonlinear approximation theory see [DeV98]. Currently, numerous efforts of this type are made in different research fields such as image processing, data compression, partial differential equations. In this monograph, the adaptation concepts for partial differential equations are of special interest which shall be briefly reviewed. A naive technique is the *remeshing* of the grid where a fixed number of mesh points is relocated. Obviously this concept is aiming at balancing the error with a fixed number of points rather than reducing the error to a given tolerance. In order to meet a fixed error tolerance the grid adaptation has to allow for *mesh enrichment*, i.e., locally refining and coarsening the mesh. This may result in an unstructured grid with locally hanging nodes. Instead of refining the grid it is also possible to increase locally the approximation order p or apply a different discretization operator for a fixed grid. This leads to a *hybrid discretization*. Of course both strategies can be combined. More details on this subject can be found for instance in [Sch98]. For time-dependent problems one might also apply *local time steps*. In this case, the constraint for the time discretization due to a CFL number is locally weakened without causing instabilities. Hence the solution may evolve faster in time for coarse cells than for fine cells. Of course, the solution has to be synchronized in case of instationary problems but not necessarily for steady state problems. For details see e.g. [BO84]. Instead of adapting the discretization one might also locally change the underlying *model*, e.g. linearize the model or neglect higher order derivatives if the corresponding physical effects are small.

Although the above techniques differ in the adaptation strategy they have one problem in common, namely, the control of the adaptation. Two strategies that are applied in the context of grid refinement shall be briefly summarized. Here we distinguish between concepts based on *error indicators* and *error estimators*, respectively. In case of error indicators, the grid is remeshed, e.g., according to steep (discretely approximated) gradients of a physically relevant quantity or other indicators. However, this strategy provides only control on the grid refining and coarsening but no information about the error of the approximation. A reliable concept is the *error–balancing strategy*. The goal is to equilibrate the error. To this end, a tolerance *tol* and a maximal number of discretization points N_{max} are fixed. By means of residual–based a posteriori estimates the grid is locally refined until a local error estimator is proportional to the ratio tol/N_{max}. This leads to an optimal mesh size distribution. In practice, it cannot be realized. Therefore one is aiming at an almost quasi equidistribution of the error tolerances. Numerous results on a posteriori error estimates have been reported in the literature for elliptic problems, see [Ver95, EEHJ95, BR96, HR02], parabolic problems [EJ91, EJ95] and hyperbolic problems see [Tad91, CCL94, Vil94, JS95, CG96, Noe96, SH97, KO99]. During the last decade new strategies have been developed based on *multiscale techniques*. Here wavelet techniques have become very popular. The basic idea is to decompose the trial space into a coarser approximation space and a complement space spanned by so–called wavelet functions. This decomposition is recursively applied to the coarse approximation space. Finally, we obtain a decomposition of the trial space into the coarsest approximation space and a sequence of complement spaces representing the difference between the approximation spaces. Performing a change of basis the solution can now be equivalently represented in terms of the single–scale basis corresponding to the trial space of the finest approximation space and the multiscale or wavelet basis, respectively. Since the coefficients of the wavelet expansion, so–called wavelet coefficients or details, may become small whenever the solution is locally smooth, data compression can be performed applying threshold techniques. For instance, one only keeps the N largest coefficients. Here the objective is to minimize the error by N coefficients (see e.g. [CDD01]). This corresponds to the idea of *best N-term approximation.* Alternatively, a tolerance ε can be fixed and all details smaller than this threshold value are discarded. Here the idea is to reduce the total number of coefficients to a small number of significant coefficients where the error to the approximate solution of the underlying approximation space is proportional to ε (see e.g. [GM99a, CKMP01]). In order to control the threshold error we need to relate coefficient norms to function norms.

The present work is concerned with developing and analyzing an adaptive finite volume scheme (FVS) for the approximation of multidimensional hyperbolic conservation laws. The concept is based on multiscale techniques which have already been mentioned above. First work on this subject has been reported by Harten [Har94, Har95]. Here the goal is the acceleration of a *given* FVS on a grid of *uniform* resolution by a *hybrid* flux computation. The core ingredient is the *multiscale decomposition* of a sequence of averages corresponding to a grid of finest resolution into a sequence of *details* and *coarse grid averages.* This decomposition is performed on a sequence of *nested* grids with decreasing resolution. It can be utilized in order to distinguish smooth regions of the flow field from regions with locally strong variations in the solution. In particular, the hybrid flux evaluation can be controlled by the decomposition, i.e., expensive upwind discretizations based on Riemann solvers are only applied near discontinuities of the solution. Elsewhere cheaper linear combinations of already computed numerical fluxes on coarser scales are used instead. These correspond to finite difference approximations. In the meantime this originally one–dimensional concept has been extended to multidimensional problems on Cartesian grids [BH97, CD01], curvilinear patches [DGM00] and triangulations [SSF00, Abg97, CDKP00].

The bottleneck of Harten's strategy is the fact that the *computational complexity*, i.e., the number of floating point operations as well as the memory requirements, corresponds to the globally finest grid. In view of multidimensional applications, this is a severe disadvantage. Recently, a real *adaptive* approach has been presented in [GM99a] and has been investigated in [CKMP01] where the computational complexity is proportional to the problem–inherent degrees of freedom. The basic idea of this concept is to determine an *adaptive grid* by means of a sequence of *truncated details.* The set of significant details can be interpreted as a *tree.* Then the adaptive grid is constructed by locally refining the grid according to the tree of *significant details.* This leads to an unstructured grid with hanging nodes. In order to restrict the computational complexity to the number of significant details the multiscale transformation is only performed on the set of significant details and the averages corresponding to the adaptive grid. It turned out that the *grading* of the tree simplifies the local transformation without increasing the complexity. In particular, the leaves of the graded tree directly correspond to the adaptive grid.

In order to preserve the accuracy of the reference FVS with respect to the finest grid the numerical fluxes on the adaptive grid have to be evaluated judiciously. No error at all is introduced when locally performing the flux evaluation by means of the averages on the *finest* scale. However, this requires a local reconstruction process by which the computational complexity is increased for multidimensional problems. Investigations for a one–dimensional scalar equation verify that for first order approximations the accuracy of the adaptive FVS is much less than that of the reference FVS (see [CKMP01]). However, parameter studies show that in case of higher order accurate FVS based on reconstruction techniques this constraint can be weakened. Here it is possible to utilize the given local averages directly instead of computing the averages on the finest scale. The target accuracy is still preserved by means of the solver–inherent reconstruction step.

A point of special interest is the reliability of the scheme, i.e., the perturbation error introduced by the truncation process can be controlled over all time levels. For this purpose analytically rigorous estimates have to be derived by which the details on the *new* time level can be estimated by those already computed in the *previous* time step. For the one–dimensional scalar case this prediction has been analytically investigated in [CKMP01]. The results derived there justify for the first time the heuristic approach suggested by Harten.

By now the new adaptive multiresolution concept has been applied by several groups with great success to different applications, e.g., 2D–steady state computations of compressible fluid flow around air wings modeled by the Euler and Navier–Stokes equations, respectively, on block–structured curvilinear grid patches [BGMH$^+$01], non–stationary shock–bubble interactions on 2D Cartesian grids for Euler equations [Mül02], backward–facing step on 2D triangulations [CKP02] and simulation of a flame ball modeled by reaction–diffusion equations on 3D Cartesian grids [RS02].

This book presents a self-contained account of the above adaptive concept for conservation laws. The main objectives are the construction and the analysis of the local multiscale transformation, the derivation of the adaptive FVS and a rigorous error analysis. New applications on Cartesian and curvilinear grids for the 2D Euler equations are presented which verify that the solver can be applied to real world problems. According to this the outline of the present work is as follows: In Chap. 1 the governing equations are presented and some of the characteristic properties are summarized. This is concluded by a brief introduction to Godunov–type schemes which form an important class of FVS frequently applied to approximate the solution of conservation laws. The multiscale setting is outlined in Chap. 2. It is based on a *hierarchy of nested grids*. As a simple but important example the Haar basis is presented to outline the basic principles and the goal of the multiscale setting. This motivates the general framework of *biorthogonal wavelets* and *stable completions.* Modifying the Haar basis appropriately leads to a new basis with "good" cancellation properties which is utilized in the adaptive scheme. In Chap. 3 the local multiscale analysis is introduced by means of the modified basis. In particular, the tree of significant details, the grading of the tree and the construction of the adaptive grid are investigated in some detail. The performance of the local multiscale transformation is analyzed in detail which results in sufficient conditions for the grading of the details. The construction of the adaptive FVS is presented in Chap. 4. In particular, several strategies for the evaluation of the numerical fluxes are discussed and the construction of the prediction set of significant details on the new time level is outlined. An error analysis is presented in Chap. 5. It is based on an ansatz originally considered by Harten [Har95] in the context of his hybrid scheme and the results derived in [CKMP01]. An efficient implementation of the adaptive scheme crucially depends on the data structures by which the algorithm is realized. This is no longer a trivial task as it is for schemes based on structured meshes. In order to realize optimal computational complexity the data structures have to be adapted judiciously to the underlying adaptive algorithm. Such appropriate data structures are discussed in Chap. 6. Finally, in Chap. 7, some relevant numerical examples illustrate the computational complexity and accuracy behavior of the scheme and problems arising in engineering applications are presented.

Acknowledgments: It is a great pleasure for me to express my gratitude to those persons who have been supporting my scientific work. In particular, I wish to thank my three mentors: first of all, Prof. Wolfgang Dahmen, RWTH Aachen, who introduced to me the world of wavelets and showed me the mathematical concepts beyond the technical details; furthermore, Prof. Josef Ballmann, RWTH Aachen, who depicted to a mathematician the physics behind the mathematical models and last but not least Prof. Rolf Jeltsch, ETH Zürich, for his enthusiasm and optimism encouraging me to start with a scientific career. Moreover, I would like to thank my colleagues at the Institut für Geometrie und Praktische Mathematik, RWTH Aachen. Among others I would like to point out Dr. K.-H. Brakhage for his neverending help concerning any kind of software related problems, Dipl.-Math. Alexander Voß for discussions on software concepts and the design of data structures and Frank Knoben for his invaluable work as system administrator. The present work was supported in parts by the collaborative research center SFB 401 "Modulation of Flow and Fluid–Structure Interaction at Airplane Wings" and the EU–TMR Network "Multiscale Methods in Numerical Simulation". The latter made it possible to spend six months at the Laboratoire d'Analyse Numérique, Université Pierre et Marie Curie, Paris VI. This research stay had a strong influence on my scientific work. In particular, I thank my collaborators Prof. Albert Cohen, Dr. Sidi M. Kaber and Dr. Marie Postel. Furthermore, I would like to thank Dipl.-Ing. Frank Bramkamp and Dipl.-

Math. Philipp Lamby for their cooperation in developing the new flow solver QUADFLOW which verifies that the present adaptive multiscale concept is a useful tool in solving efficiently and reliably real world problems. Prof. Wolfgang Dahmen and Prof. Sebastian Noelle, RWTH Aachen, and Prof. Thomas Sonar, TU Braunschweig, acted as referees for my habilitation thesis, and I would like to thank them for their careful reading of my work. The constructive comments of several unknown referees contributed significantly to the final version of the present book. I would also like to thank Dr. Martin Peters and Thanh-Ha Le Thi, Springer Verlag, for the professional and pleasant cooperation. Last but not least I would like to express my deepest gratitude to my wife and colleague Dr. Birgit Gottschlich-Müller for her collaboration and her love. She always encouraged me to continue with my work.

Aachen, September 2002 *Siegfried Müller*

Table of Contents

1 Model Problem and Its Discretization

The objective of this chapter is the introduction of conservation laws and how to approximate their solution. Basic aspects concerning existence and uniqueness are briefly reviewed. Motivated by analytical considerations finite volume schemes based on Godunov–type methods are summarized which are frequently applied in numerical simulations.

1.1 Conservation Laws

The present thesis is concerned with evolution equations of the form

$$\frac{\partial\, \mathbf{u}(t,\mathbf{x})}{\partial\, t} + \sum_{i=1}^{d} \frac{\partial\, \mathbf{f}_i(\mathbf{u}(t,\mathbf{x}))}{\partial\, x_i} = \mathbf{0}, \qquad (t,\mathbf{x}) \in (0,T) \times \Omega \tag{1.1}$$

describing the temporal change of the *conservative quantities* $\mathbf{u} : [0,T] \times \overline{\Omega} \rightarrow \mathcal{D}$ due to spatial variations quantified by the *fluxes* $\mathbf{f}_i : \mathcal{D} \rightarrow \mathrm{R}^m$, $i = 1, \ldots, d$, for each of the spatial directions. Here the open set $\Omega \subset \mathrm{R}^d$ denotes the *computational domain* and $\mathcal{D} \subset \mathrm{R}^m$ the *space of admissible states.* In the literature, evolution equations of the type (1.1) are referred to as scalar ($m = 1$) or system ($m > 1$) of *conservation laws.* They arise from the principles of balancing mass, momentum and energy in classical continuum mechanics under local regularity assumptions. In order to admit physically meaningful discontinuities, it is more convenient to consider the evolution of an averaged conservative quantity. By the divergence theorem, the partial differential equation (1.1) can be transformed into the ordinary differential equation

$$\frac{d}{dt} \int_V \mathbf{u}(t,\mathbf{x})\, d\mathbf{x} = - \int_{\partial V} \mathbf{f}_{\mathbf{n}(\mathbf{x})}(\mathbf{u}(t,\mathbf{x}))\, d\mathbf{x}, \tag{1.2}$$

where V is any (time–independent) control volume with Lipschitz continuous boundary and $\mathbf{n}$ denotes the outer normal to ∂V. The *flux in normal direction* is defined by

$$\mathbf{f}_{\mathbf{n}}(\mathbf{u}) := \sum_{i=1}^{d} \mathbf{f}_i(\mathbf{u})\, n_i.$$

Obviously, the quantity $E_V(t) := \int_V \mathbf{u}(t,\mathbf{x})\,d\mathbf{x}$ remains constant if the right hand side in (1.2) vanishes, i.e., the incoming and outgoing information across the boundary ∂V are balanced. Note, that the evolution equation (1.2) corresponds to the balance equations in continuum mechanics. Later, this equation will be the starting point for the design of an appropriate numerical scheme.

Throughout this work the fluxes are assumed to be smooth, i.e., $\mathbf{f}_i \in C^2(\mathcal{D}, \mathsf{R}^m)$ and the Jacobian of the normal flux

$$\mathsf{A}(\mathbf{u},\mathbf{n}) := \frac{\partial\,\mathbf{f}_{\mathbf{n}}(\mathbf{u})}{\partial\,\mathbf{u}} = \sum_{i=1}^{d} \frac{\partial\,\mathbf{f}_i(\mathbf{u})}{\partial\,\mathbf{u}} n_i \tag{1.3}$$

has m real eigenvalues $\lambda_i(\mathbf{u},\mathbf{n})$ as well as m linearly independent right eigenvectors $\mathbf{r}_i(\mathbf{u},\mathbf{n})$ and left eigenvectors $\mathbf{l}_i(\mathbf{u},\mathbf{n})$, $i = 1,\ldots,m$, respectively, for all $\mathbf{u} \in \mathcal{D}$ and $\mathbf{n} \in \mathsf{R}^d$ with $\|\mathbf{n}\|_2 = 1$. In this case the system is called *hyperbolic*. Some examples shall be presented in the following.

Example 1. (Traffic Flow ($m = d = 1$), [LeV92, Krö97])
Here u denotes the traffic density, i.e., cars per unit length, and $f(u)$ the traffic flow, i.e., cars per unit time. A simple model for the flow is given by

$$f(u) = -(u - u_m)^2 + f_0$$

where the flow increases until a maximal speed u_m, e.g. speed limit, is reached and then decreases.

Example 2. (Buckley–Leverett equation ($m = d = 1$), [LeV92, Krö97])
This is a simple model for two phase fluid flow, e.g. water and oil, in porous media such as sand, which arises in oil recovery when water is pumped into an oil field forcing out the oil. Here u is the water saturation and the relative permeability is given by

$$f(u) = \frac{u^2}{u^2 + a(1-u^2)},$$

where a is the ratio of the viscosity coefficients corresponding to the two phases.

Example 3. (Euler equations ($m = d + 2,\ d \geq 1$), [CF48, CM79])
These equations are derived in gas dynamics taking only inviscid effects into account. The motion of the fluid is described by the continuity equation, the momentum equation and the energy equation for conservation of mass, momentum and energy. This results in a system of conservation laws. Here $\mathbf{u} = (\varrho, \mathbf{m}, \varrho\,E)$ is the vector of mass density ϱ, momentum $\mathbf{m} \in \mathsf{R}^d$ and the total energy per unit volume $\varrho\,E$. The corresponding fluxes can be written in the form

$$\mathbf{f}_i(\mathbf{u}) = \begin{pmatrix} m_i \\ m_i\mathbf{m}/\varrho + p\mathbf{e}_i \\ m_i(\varrho\,E + p)/\varrho \end{pmatrix}$$

where $\mathbf{e}_i \in \mathsf{R}^d$ denotes the ith unit vector in R^d. Obviously, the resulting system has to be completed by an additional equation for the pressure p, the so-called *equation of state* $p = p(\varrho, \mathbf{m}, \varrho E)$. Two frequently used models are

$$\text{polytropic gas:} \quad p = (\gamma - 1)\left(\varrho E - 0.5\,\mathbf{m}^2/\varrho\right), \; \gamma > 1$$
$$\text{isentropic gas:} \quad p = \kappa \varrho^\gamma, \; \kappa > 0$$

To ensure a unique solution initial conditions and in case of a bounded domain Ω boundary conditions have to be imposed, i.e.,

$$\mathbf{u}(0, \mathbf{x}) = \mathbf{u}_0(\mathbf{x}), \qquad \mathbf{x} \in \Omega, \tag{1.4}$$
$$\mathbf{u}(t, \mathbf{x}) = \mathbf{u}_\Gamma(t, \mathbf{x}), \quad \mathbf{x} \in \Gamma(t) \subset \partial\Omega, \; t \in (0, T). \tag{1.5}$$

In the sequel the initial value problem (1.1), (1.4) is referred to as IVP and the initial boundary value problem (1.1), (1.4), (1.5) as IBVP. Of course, the conditions (1.4) and (1.5) have to be chosen judiciously in agreement with existence and uniqueness results for the IVP and the IBVP, respectively. The number of boundary conditions μ depends on the attached flow field. In particular, at most μ conditions are admissible if there exist μ negative eigenvalues of $\mathsf{A}(\overline{\mathbf{u}}, \mathbf{n})$ where $\overline{\mathbf{u}} = \lim_{h \to 0+} \mathbf{u}(t, \mathbf{x} - h\,\mathbf{n})$ denotes the limit of the solution $\mathbf{u}$ at the boundary point $\mathbf{x} \in \partial\Omega$ and $\mathbf{n}$ the corresponding outer unit normal. Note, that boundary conditions are only admissible at *inflow* boundaries ($\mu > 0$) but not at *outflow* boundaries ($\mu = 0$). For more details on initial and boundary conditions the reader is referred to [Kre70, KL89].

In the following the question of existence and uniqueness is addressed in order to emphasize the problems arising later when (1.1) is discretized. A first result shows that a differentiable or *classical* solution in general only exists for small time even if the initial data are smooth.

Theorem 1. *(Local Existence of classical solution)*
Let $m = 1$, $d \geq 1$ and $\Omega = \mathsf{R}^d$. Let $f_i \in \mathsf{C}^2(\mathsf{R}, \mathsf{R})$, $i = 1, \ldots, d$, and $u_0 \in C^1(\mathsf{R}^d, \mathsf{R})$ be bounded such that

$$\mu_{i,j} := \inf_{\mathbf{x} \in \mathsf{R}^d} \left\{ f_i''(u_0(\mathbf{x})) \frac{\partial\, u_0(\mathbf{x})}{\partial\, x_j} \right\} > -\infty, \; i, j = 1, \ldots, d.$$

Then there exists $\delta > 0$ such that a classical solution of the IVP exists for all $t \in [0, \delta)$ given by $u(t, \mathbf{x}) = u_0(\Psi_t^{[-1]}(\mathbf{x}))$ with $\Psi_t(\mathbf{x}_0) := \mathbf{x}_0 + t\,\mathbf{f}'(u_0(\mathbf{x}_0))$ and $\mathbf{f} = (f_1, \ldots, f_d)^T$.

In the one-dimensional case δ is determined by $\delta = -1/\mu$ ($\mu \equiv \mu_{1,1} < 0$) and $\delta = \infty$ ($\mu \geq 0$) elsewhere.

Sketch of proof: The proof is based on the observation that the classical solution has to be constant along *characteristic* lines determined by the system of ordinary differential equations

$$\frac{d\,\mathbf{x}(t)}{dt} = \mathbf{f}'(u(t, \mathbf{x}(t))), \quad \mathbf{x}(0) = \mathbf{x}_0.$$

The basic idea is now to determine for each point $(t, \mathbf{x}) \in (0, \delta) \times \mathsf{R}^d$ a unique characteristic line emanating from a point $\mathbf{x}_0$ on the plane $t = 0$. Verifying invertibility of the map Ψ_t leads to the assertion. □

For results on systems of conservation laws see [FM72, FL71].

From a physical point of view solutions of (1.1) exhibiting discontinuities are meaningful. In order to cover such cases the notion of the solution has to be generalized appropriately.

Definition 1. *(Weak solution)*
Let $\mathbf{u}_0 \in L^\infty_{loc}(\Omega, \mathcal{D})$. Then $\mathbf{u} \in L^\infty_{loc}([0,T] \times \Omega, \mathcal{D})$ is called a weak solution *of (1.1) subject to the initial and boundary conditions (1.4,1.5) if and only if the equation*

$$\int_0^T \int_\Omega \mathbf{u}\frac{\partial \varphi}{\partial t} + \sum_{i=1}^d \mathbf{f}_i(\mathbf{u})\frac{\partial \varphi}{\partial x_i}\, d\mathbf{x}\, dt + \int_\Omega \mathbf{u}_0(\mathbf{x})\varphi(0,\mathbf{x})\, d\mathbf{x} = \int_0^T \int_{\partial\Omega} \sum_{i=1}^d \mathbf{f}_i(\mathbf{u})\, n_i\, \varphi\, d\mathbf{x}\, dt \tag{1.6}$$

holds for all test functions $\varphi \in C_0^\infty(\mathsf{R}^{d+1})$ where $\mathbf{n}$ denotes the outer unit normal to $\partial\Omega$. In case of an IVP the right hand side vanishes.

Note, that equation (1.6) corresponds in a natural way to the integral equations, so–called balance equations, derived in continuum mechanics. In this setting the conservation laws (1.1) can be viewed as local balance equations provided local regularity assumptions hold.

Obviously, weak solutions are not necessarily smooth. Hence, discontinuities are now admissible. However, the discontinuities are not arbitrary but have to satisfy the well–known Rankine–Hugoniot conditions which are presented in the following theorem for general multidimensional systems of conservation laws.

Theorem 2. *(Rankine–Hugoniot conditions)*
Let $\mathbf{u} \in L^\infty_{loc}([0,T] \times \Omega, \mathsf{R}^m)$ be a piecewise smooth weak solution of (1.1) and $\Gamma^* := \{(t,\mathbf{x}) \,:\, g(t,\mathbf{x}) = 0\}$ a discontinuity surface which is implicitly given by the function $g : [0,T] \times \Omega \to \mathsf{R}$. The normal vector of Γ^* can be represented by

$$\mathbf{n}^*(t,\mathbf{x}) = \mathbf{n}^* = (n_0, \mathbf{n}_d)^T = (\partial_t g, \partial_{x_1} g, \ldots, \partial_{x_d} g)^T. \tag{1.7}$$

Then for the limiting values $\mathbf{u}_\pm(t,\mathbf{x}) := \lim_{\delta\to 0\pm} \mathbf{u}(t, \mathbf{x} + \delta\, \mathbf{n}(t,\mathbf{x}))$ the relation

$$-\sum_{i=1}^d n_i(\mathbf{f}_i(\mathbf{u}_+) - \mathbf{f}_i(\mathbf{u}_-)) = n_0(\mathbf{u}_+ - \mathbf{u}_-) \tag{1.8}$$

holds.

Sketch of proof: (See also [Liu76].) Consider a neighborhood $D \subset (0,T)\times\Omega$ of a point $(t,\mathbf{x}) \in \Gamma^*$ which is split into two parts by the discontinuity surface. In each of the two subdomains the solution $\mathbf{u}$ is smooth according to the assumptions. Therefore the partial derivatives in the integral equation (1.6) can be shifted on each subdomain from the test function φ to $\mathbf{u}$ and $\mathbf{f}_i$, respectively, by means of integration by parts. Due to the evolution equation (1.1) only the boundary integrals corresponding to the discontinuity surface do not vanish. Then the assertion immediately follows for meas$(D) \to 0$ and incorporating (1.7). □

Unfortunately, the Rankine–Hugoniot conditions do not ensure uniqueness of the IVP, i.e., (1.8) is only a necessary condition. As is verified in several textbooks, e.g. [LeV92, Krö97], it is possible to construct different weak solutions of the IVP. In order to characterize a "physically meaningful" solution additional constraints have to be imposed. In the literature, there exists a variety of so–called *entropy concepts* motivated by the entropy in thermodynamics by which a unique solution can be characterized in some cases. For scalar equations Kruzhkov's concept seems to be the most general concept and shall be presented here.

Definition 2. *(Kruzhkov's entropy condition [Kru70])*
A weak solution is called an entropy solution *if and only if the inequality*

$$\int_0^T \int_\Omega |u-k| \frac{\partial \varphi}{\partial t} + \mathrm{sign}(u-k) \sum_{i=1}^d |f_i(u) - k| \frac{\partial \varphi}{\partial x_i} \, d\mathbf{x}\, dt + \\ \int_0^T \int_{\partial\Omega} \mathrm{sign}(k) \sum_{i=1}^d |f_i(\gamma(u(t,\mathbf{x}))) - f_i(k)| \, n_i \varphi \, d\mathbf{x}\, dt \geq 0 \qquad (1.9)$$

holds for all nonnegative test functions $\varphi \in C_0^\infty(\mathsf{R}^+ \times \mathsf{R}^d)$ and all $k \in \mathsf{R}$. Here $\mathbf{n}$ denotes the outer unit normal to $\partial\Omega$ and $\gamma : BV((0,T)\times\Omega) \to L^\infty(\{0\}\times\Omega)\times L^\infty((0,T)\times\partial\Omega)$ the trace map according to [BBB73, BLN79]. In case of an IVP the boundary integral vanishes.

Note, that in the scalar case the known entropy concepts are equivalent. In particular, the Kruzhkov entropy criterion coincides with the entropy pair concept, because the cone of convex functions is spanned by $|x-y|$.

By means of (1.9) Kruzhkov proved that the entropy solution of the IVP is unique [Kru70]. The existence was proved by Kuznetsov [Kuz67]. For the IBVP existence and uniqueness were also established [BLN79, Krö97].

For systems of conservation laws there exist by now several entropy concepts but they are not as mature as in the scalar case. For a discussion on this subject the reader is referred to Dafermos' book [Daf00] where several concepts are discussed in the context of continuum mechanics.

1.2 Finite Volume Methods

In general a closed form analytical representation of the solution of the IVP and the IBVP is not available. In order to determine an approximate solution several strategies have been investigated. A naive discretization is based on finite differences. However, in the presence of discontinuities they produce oscillations which cause a breakdown of the computation. These schemes can be stabilized by adding *artificial viscosity* which leads to a significant loss in accuracy. In particular, the shape of discontinuities is smeared. Nevertheless *finite difference schemes* are frequently applied in industry, e.g. aviation industry, since their implementation is easy and they are very efficient due to their simple structure which is supported by modern computer hardware, in particular vector architectures. Important representatives for finite difference schemes are the Lax–Friedrichs scheme [Lax54] which became a helpful tool in proving existence and uniqueness of conservation laws and the Lax–Wendroff scheme [LW60] as a typical higher order discretization.

Another strategy is based on *finite element methods.* Once again, standard Galerkin discretizations usually applied to elliptic problems are not stable when applied to conservation laws. These schemes also have to be stabilized by additional viscosity, see for instance streamline–diffusion methods [Joh87]. In recent years a new approach has been derived based on discontinuous, piecewise polynomial functions, so–called discontinuous Galerkin discretizations [CKS99]. This approach can be viewed as a combination of finite element methods and finite volume methods.

A natural choice for the discretization of conservation laws are finite volume methods. Here Godunov's pioneering work [God59] has to be mentioned as the prototype of modern FVS. The principle idea is to evolve an approximate solution represented by piecewise constant data in time by solving a so–called *Riemann problem* at the interfaces where the approximate solution jumps. Since then Godunov–type schemes have been studied extensively. Two main streams of research shall be mentioned. One is concerned with the solution of the Riemann problem and its approximation, see Toro's book [Tor97] for a survey. Another subject is the development of high–order accurate schemes based on *reconstruction techniques*, see e.g. [HEOC87]. Since the present work is concerned with the acceleration of FVS the structure of Godunov–type schemes will be briefly motivated and the core ingredients are outlined.

In the context of FVM the natural discretization is by cell averages

$$\hat{\mathbf{u}}_j^n := \frac{1}{|V_j|}\int_{t_n}^{t_{n+1}} \mathbf{u}(t,\mathbf{x})\,d\mathbf{x}\,dt, \quad |V_j| := \int_{V_j} 1\,d\mathbf{x}$$

where the computational domain Ω is partitioned into cells V_k, i.e., $\Omega = \bigcup_k V_k$. The cells V_k are always assumed to be closed. They are typically intervals in 1D, triangles and quadrangles in 2D, tetrahedra and hexahedra in 3D, and the time interval $[0,T]$ is decomposed by a temporal mesh

$0 = t_0 < \ldots < t_{N_T} = T$. Here the temporal partition is assumed to be uniform, i.e., $t_{n+1} - t_n = \tau$ for all $n = 0, \ldots, N_T - 1$. This is no constraint for the construction of the adaptive scheme but simplifies the notation, because *spatial* mesh adaptation is only considered here. For these averages the evolution equation

$$\hat{\mathbf{u}}_k^{n+1} = \hat{\mathbf{u}}_k^n - \frac{1}{|V_k|} \int_{t_n}^{t_{n+1}} \int_{\partial V_k} \mathbf{f}_{\mathbf{n}_k(\mathbf{x})}(\mathbf{u}(t, \mathbf{x}))\, d\mathbf{x}\, dt \tag{1.10}$$

is derived from (1.2). According to the neighboring cells V_l of V_k the boundary ∂V_k can be decomposed into different parts $\Gamma_{k,l}$, i.e.,

$$\partial V_k = \bigcup_l \Gamma_{k,l} \qquad \text{with} \quad \Gamma_{k,l} := \partial V_k \cap \partial V_l .$$

In case of a bounded domain Ω a layer of ghost cells outside the domain is introduced which is attached to the boundary $\partial \Omega$. Then equation (1.10) can be rewritten as

$$\hat{\mathbf{u}}_k^{n+1} = \hat{\mathbf{u}}_k^n - \frac{\tau}{|V_k|} \sum_l |\Gamma_{k,l}|\, \hat{\mathbf{f}}_{k,l}^n, \quad |\Gamma_{k,l}| := \int_{\Gamma_{k,l}} 1\, d\mathbf{x}. \tag{1.11}$$

Since in the one-dimensional case, the edges $\Gamma_{k,l}$ coincide with a single grid point of the mesh, the value of $|\Gamma_{k,l}|$ is by definition put to 1. Here the *average flux* across the interface is defined by

$$\hat{\mathbf{f}}_{k,l}^n := \frac{1}{\tau} \frac{1}{|\Gamma_{k,l}|} \int_{t_n}^{t_{n+1}} \int_{\Gamma_{k,l}} \mathbf{f}_{\mathbf{n}_{k,l}(\mathbf{x})}(\mathbf{u}(t, \mathbf{x}))\, d\mathbf{x}\, dt. \tag{1.12}$$

Note, that the outer normals $\mathbf{n}_{k,l}$ and $\mathbf{n}_{l,k}$ corresponding to the cells V_k and V_l, respectively, differ only in sign. Therefore the average fluxes satisfy the *conservation property*

$$\hat{\mathbf{f}}_{k,l}^n = -\hat{\mathbf{f}}_{l,k}^n .$$

Substituting in (1.11) the averages $\hat{\mathbf{u}}_j^n$ and the average fluxes $\hat{\mathbf{f}}_{k,l}^n$ by some approximation $\mathbf{v}_k^n$ and $\mathsf{F}_{k,l}^n$, respectively, leads to the general representation of a FVS

$$\mathbf{v}_k^{n+1} = \mathbf{v}_k^n - \frac{\tau}{|V_k|} \sum_l |\Gamma_{k,l}|\, \mathsf{F}_{k,l}^n .$$

Of course, the key ingredient is an appropriate computation of the *numerical fluxes* $\mathsf{F}_{k,l}^n$. To obtain a robust and high–order accurate scheme three basic requirements arise, namely, *consistency*, *conservation* and *accuracy*, which have to be taken into account in the construction of the numerical fluxes

- $\mathbf{v}_m^n \equiv \mathbf{v} \;\Rightarrow\; \mathsf{F}_{k,l}^n = \frac{1}{|\Gamma_{k,l}|} \int_{\Gamma_{k,l}} \mathbf{f}_{\mathbf{n}_{k,l}(\mathbf{x})}(\mathbf{v})\, d\mathbf{x}$,
- $\mathsf{F}_{k,l}^n = -\mathsf{F}_{l,k}^n$,

– $\mathsf{F}^n_{k,l} = \hat{\mathbf{f}}^n_{k,l} + \mathcal{O}(h^q) + \mathcal{O}(\tau^p)$,

where h denotes the largest diameter of the cells involved in the discretization stencil of the numerical flux $\mathsf{F}^n_{k,l}$. An appropriate construction is presented in the following. It consists of four basic steps:

Step 1: Quadrature of Spatial Boundary Integral. The integral over the boundary part $\Gamma_{k,l}$ in (1.12) is approximated by a quadrature (d=2) or a cubature (d=3) formula of order N_Q with P_Q quadrature points $\mathbf{x}_i^{k,l}$ and weights $\omega_i^{k,l}$, i.e.,

$$\int_{\Gamma_{k,l}} \mathbf{f}_{\mathbf{n}_{k,l}(\mathbf{x})}(\mathbf{u}(t,\mathbf{x}))\, d\mathbf{x} = \sum_{i=1}^{P_Q} \omega_i^{k,l} \mathbf{f}_{\mathbf{n}_{k,l}(\mathbf{x}_i^{k,l})}(\mathbf{u}(t,\mathbf{x}_i^{k,l})) + \mathcal{O}(h^{N_Q}).$$

In the one–dimensional case ($d=1$) this step is void since $\Gamma_{k,l}$ reduces to a single point.

Step 2: Flux approximation. The fluxes in normal direction have to be evaluated at the quadrature points where the solution $\mathbf{u}$ is not yet known. Therefore the fluxes are substituted by approximations of the form $\mathsf{G}(\mathbf{v},\mathbf{w},\mathbf{n})$ that are assumed to be consistent in the sense that $\mathsf{G}(\mathbf{v},\mathbf{v},\mathbf{n}) = \mathbf{f}_{\mathbf{n}}(\mathbf{v})$. An appropriate flux approximation is the *Godunov flux* introduced in [God59]. This approximation is based on the solution of the one–dimensional Riemann problem

$$\begin{aligned} \frac{\partial\, \mathbf{u}(t,x)}{\partial\, t} + \frac{\partial\, \mathbf{f}_{\mathbf{n}}(\mathbf{u}(t,x))}{\partial\, x} &= \mathbf{0}, \\ \mathbf{u}(0,x) &= \begin{cases} \mathbf{u}_L \ , \ x<0, \\ \mathbf{u}_R \ , \ x>0, \end{cases} . \end{aligned} \tag{1.13}$$

The solution of this problem is known to be self–similar, i.e., it is constant along rays $x/t = const$, and can be represented as $\mathbf{u}(t,x) = \mathsf{U}(\mathbf{u}_L,\mathbf{u}_R,x/t)$. For details on the existence and the construction of the Riemann solution see [Gel59] (scalar) and [Liu75] (system). Then the Godunov flux is determined by $\mathsf{U}(\mathbf{v}_k,\mathbf{v}_l,0)$ where the initial values are the averages on both sides of the interface $\Gamma_{k,l}$. Obviously, this leads to a first order approximation

$$\mathsf{G}(\mathbf{v}_k,\mathbf{v}_l,\mathbf{n}_{k,l}) := \mathsf{U}(\mathbf{v}_k,\mathbf{v}_l,0) = \mathbf{f}_{\mathbf{n}_{k,l}(\mathbf{x}_i^{k,l})}(\mathbf{u}(t,\mathbf{x}_i^{k,l})) + \mathcal{O}(h),$$

since the exact averages $\mathbf{u}_k$ and $\mathbf{u}_l$ are only first order approximations of the point value $\mathbf{u}(t,\mathbf{x}_i^{k,l})$. If the solution u would be known, then no error is introduced when evaluating the numerical fluxes at $\mathbf{v} = \mathbf{w} = \mathbf{u}(t,\mathbf{x}_i^{k,l})$ and $\mathbf{n} = \mathbf{n}_{k,l}(\mathbf{x}_i^{k,l})$ according to the consistency of the flux approximation.

Solving such local Riemann problems accounts for a substantial part of the overall computational effort. A less expensive approximation is based on linearizing the Riemann problem in the sense that the flux $\mathbf{f}_{\mathbf{n}}$ is replaced by the linear approximation $\overline{\mathbf{f}}_{\mathbf{n}}(\mathbf{u}) := \mathsf{A}(\overline{\mathbf{u}}, \mathbf{n})\, \mathbf{u}$ with respect to an intermediate state $\overline{\mathbf{u}}$. This was first introduced by Roe [Roe81]. For more details on approximate Riemann solvers see [Tor97].

Step 3: Reconstruction. The low accuracy of the Godunov and Roe flux approximation can be remedied by means of reconstruction techniques introduced by Harten et. al [HEOC87]. The basic idea is to construct an approximation to a scalar function $w : \Omega \to \mathsf{R}$ based on its averages $\{\hat{w}_k\}_k$ such that the corresponding *reconstruction function* $R^{N_R}(\cdot, w)$ has a desired *discretization error*, satisfies the *conservation property* and is *essentially non–oscillatory* (ENO)

- $R^{N_R}(\mathbf{x}, w) = w(\mathbf{x}) + \mathcal{O}(h^{N_R}), \quad w$ sufficiently smooth,
- $\frac{1}{|V_k|} \int_{V_k} R^{N_R}(\mathbf{x}, w)\, d\mathbf{x} = \hat{w}_k,$
- $\mathbf{TV}(R^{N_R}(\cdot, w)) \le \mathbf{TV}(w) + \mathcal{O}(h^\alpha), \quad \alpha > 1.$

It should be emphasized that the reconstruction function is not necessarily smooth on the whole computational domain Ω but may jump, for instance, at the interfaces of the cells V_k. In practice, this does in fact happen, since for each cell a local reconstruction function $R_k^{N_R}(\cdot, w)$ is determined by means of the averages in the local neighborhood. In view of the ENO property the reconstruction stencil has to be chosen judiciously employing data dependent possibly unsymmetric stencils. For the 1D case an efficient strategy was developed in [HEOC87] by successively increasing the reconstruction stencil. For the multidimensional case this turns out to be not feasible. Here a fixed stencil is in general chosen where the reconstruction conditions are appropriately weighted when solving a least–squares problem (see e.g. [HC91, Abg91]). For more details on reconstruction techniques, see also [Son95].

The reconstruction concept has originally been developed for scalar functions. It can easily be extended to systems by applying the reconstruction componentwise or vectorwise. More sophisticated techniques are based on the reconstruction of the characteristic variables $w_i := \langle \mathbf{l}_i(\mathbf{u}, \mathbf{n}), \mathbf{u}\rangle,\ i = 1, \ldots, m,$ where $\mathbf{l}_i(\mathbf{u}, \mathbf{n})$ denotes the left eigenvector of the Jacobian $\mathsf{A}(\mathbf{u}, \mathbf{n})$.

Now the approximation order of the flux approximation can be increased. To this end, the averages are replaced by point values determined by the local reconstruction functions $\mathsf{R}_k^{N_R}(\cdot, \mathbf{u}(t, .))$, i.e.,

$$\mathsf{G}(\mathsf{R}_k^{N_R}(\mathbf{x}_i^{k,l}, \mathbf{u}(t, \cdot)), \mathsf{R}_l^{N_R}(\mathbf{x}_i^{k,l}, \mathbf{u}(t, \cdot)), \mathbf{n}_{k,l}) = \\ \mathbf{f}_{\mathbf{n}_{k,l}(\mathbf{x}_i^{k,l})}(\mathbf{u}(t, \mathbf{x}_i^{k,l})) + \mathcal{O}(h^{N_R}).$$

Note, that for a first order reconstruction the point values coincide with the averages according to the conservation property of the reconstruction.

Step 4: Quadrature of time integral. Finally the time integration has to be resolved. One possibility is to apply a Runge–Kutta scheme to the evolution equation (1.2), incorporating the approximations due to the preceding steps. This requires averages on some intermediate time levels which all have to be stored in the computation of one time step. In order to avoid this additional amount of storage one can apply a different strategy. For this purpose, the time integration is approximated by means of a quadrature formula, e.g. by a Gauss quadrature rule, where the fluxes have to be approximated at the Gaussian points $t_n \leq t_1^n < \ldots < t_{N_G}^n \leq t_{n+1}$. Since the solution u is not known at time level t_m^n, an approximation of $\mathbf{u}(t_m^n, \mathbf{x}_i^{k,l})$ is determined where u is expanded in a Taylor series at the point $(t_n, \hat{\mathbf{x}}_k)$. Here $\hat{\mathbf{x}}_k$ denotes the centroid of the cell V_k. Then the time derivatives are successively substituted by spatial derivatives exploiting the evolution equation (1.1). At time level t_n the averages are known. Hence the spatial derivatives can be approximated by the reconstruction function due to the known averages.

Obviously, the computation of the numerical fluxes is very expensive. Therefore one is interested in reducing the cost. Since upwind techniques are only essential close to discontinuities, cheaper finite difference approximations are sufficient in the smooth part of the solution. This requires a tool by which the data representing the flow field can be analyzed. In this regard Harten introduced a sophisticated strategy based on multiscale decompositions [Har95].

2 Multiscale Setting

The core ingredient of the adaptive FVS is the multiscale setting by which the data at hand, here cell averages, can be analyzed. In the literature, at least two settings are known which can be applied for this purpose. One framework was introduced by A. Harten [Har93a, Har93b, Har96] and extended later by [ADH98, ADH99]. Here the construction of a multiscale analysis is based on discrete data only and employs reconstruction and prediction techniques. In view of stability investigations, function spaces have to be introduced in order to benefit from functional analytic arguments. Therefore we prefer a different methodology by W. Dahmen [Dah94, Dah95, Dah96] and collaborators [CDP96] which is based on biorthogonal wavelets and stable completions. Although the settings are different, the resulting transformations can be rewritten in terms of the other setting, see [GM99b]. In the appendix of this book, we provide a short summary on wavelet theory for readers who are not familiar with wavelets at all, see Sect. B.

In the following we present a self–contained multiscale setting employed in the construction of adaptive multiscale FVS. In Section 2.1 we introduce a hierarchy of nested grids. By means of the box function and the box wavelet we will motivate that the cell averages relative to the nested grids are naturally related to biorthogonal wavelets, see Section 2.2. We will explain in Section 2.5 how to modify the box wavelets such that the new wavelets have improved cancellation properties. For this purpose we present a systematic fashion how to construct the modified box wavelets. This is related to stable completions, see Section 2.4. Finally, we determine a multiscale decomposition of the cell averages related to a finest resolution level that will be utilized in Chapter 3 to construct the adaptive grid.

2.1 Hierarchy of Meshes

As has been motivated in the previous chapter FVS are naturally related to cell averages of the solution. In order to detect singularities of the solution by means of the array of averages, we consider the difference of averages corresponding to different resolution levels. For this purpose we introduce a hierarchy of nested grids.

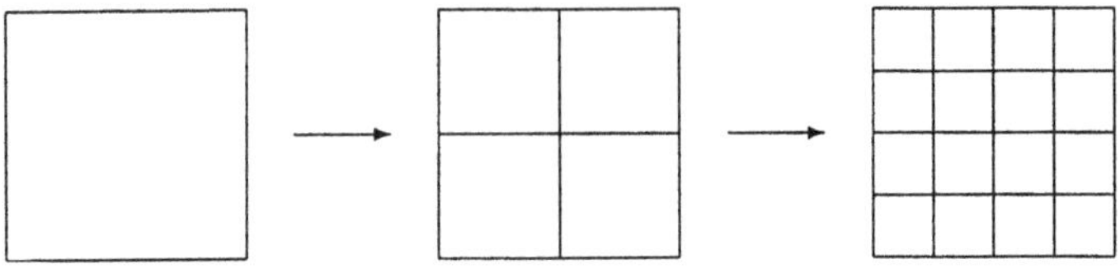

Fig. 2.1. Sequence of nested grids

Definition 3. *(Nested grid hierarchy)*
A sequence of grids $\mathcal{G}_j := \{V_{j,k}\}_{k \in I_j}$, $j = 0, \ldots, L$, *is called a* nested grid hierarchy *if the following conditions hold*

- $\Omega = \bigcup_{k \in I_j} V_{j,k}$ *with* *(partition)*

  $|V_{j,k} \cap V_{j,k'}| = 0$ *for* $k \neq k'$, $k, k' \in I_j$,
- $V_{j,k} = \bigcup_{r \in \mathcal{M}_{j,k}} V_{j+1,r}$, $k \in I_j$. *(refinement)*

Note, that the cells are always assumed to be closed.

Here the coarsest grid is indicated by 0 and the finest grid by L . Furthermore the index sets $\mathcal{M}_{j,k}$ correspond to the new cells on level $j+1$ resulting from the refinement of the cell $V_{j,k}$ which is always assumed to be closed. A simple example is shown in Fig. 2.1 where a coarse grid is successively refined with increasing refinement level. From the conditions of the nested grid hierarchy we immediately conclude that the following properties hold for the refinement sets $\mathcal{M}_{j,k}$:

- $\mathcal{M}_{j,k} \cap \mathcal{M}_{j,k'} = \emptyset$ for $k \neq k'$, $k, k' \in I_j$, (redundancy–free)
- $\bigcup_{k \in I_j} \mathcal{M}_{j,k} = I_{j+1}$. (gap–free)

For the numerical experiments performed in Chapter 7 only structured grids are considered. However, the framework presented here can also be applied to unstructured grids and irregular grid refinements. For reasons of simplicity and stability only uniform refinements are considered here, i.e.,

$$\# \mathcal{M}_{j,k} = M_r = const.$$

Relative to the grids $\mathcal{G}_j$ we introduce the so–called *box function*

$$\tilde{\varphi}_{j,k}(\mathbf{x}) := \frac{1}{|V_{j,k}|} \chi_{V_{j,k}}(\mathbf{x}) = \begin{cases} |V_{j,k}|^{-1} & , \mathbf{x} \in V_{j,k} \\ 0 & , \mathbf{x} \notin V_{j,k} \end{cases} \tag{2.1}$$

defined as the L^1–scaled characteristic function with respect to $V_{j,k}$, i.e., $\|\tilde{\varphi}_{j,k}\|_{L^1(\Omega)} = 1$. Obviously, the functions corresponding to the same discretization level are linearly independent. The nestedness of the grids as well as the linearity of integration imply the two–scale relation

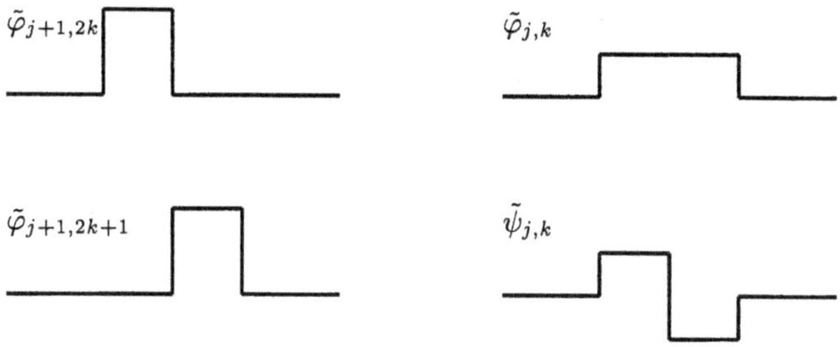

Fig. 2.2. Box function and box wavelet

$$\tilde{\varphi}_{j,k} = \sum_{r \in \mathcal{M}_{j,k}} \frac{|V_{j+1,r}|}{|V_{j,k}|} \tilde{\varphi}_{j+1,r}, \tag{2.2}$$

i.e., the coarse grid box function can be represented as a linear combination of the corresponding fine grid box functions. Consequently, the box functions can successively be computed starting with the finest level.

The reason for introducing the box functions is motivated by the fact that the averages of a scalar, integrable function $u \in L^1(\Omega)$ can be interpreted as an inner product

$$\hat{u}_{j,k} = \langle u, \tilde{\varphi}_{j,k} \rangle_\Omega \tag{2.3}$$

with the box function where the inner product is defined by

$$\langle u, v \rangle_\Omega := \int_\Omega u\, v\, d\mathbf{x}.$$

Obviously, the averages of two discretization levels are related by

$$\hat{u}_{j,k} = \sum_{r \in \mathcal{M}_{j,k}} \frac{|V_{j+1,r}|}{|V_{j,k}|} \hat{u}_{j+1,r}. \tag{2.4}$$

The goal is to transform these data into a different format of cell averages corresponding to a sequence of resolution levels. This will be motivated by a simple univariate example.

2.2 Motivation

We now consider the unit interval $\Omega = [0,1]$ where the grid hierarchy is determined by a uniform dyadic partition of $[0,1]$, i.e., $V_{j,k} = 2^{-j}[k, k+1]$, $k \in I_j := \{0, \ldots, 2^j - 1\}$. Note, that the refinement sets are $\mathcal{M}_{j,k} = \{2k, 2k+1\}$. Then the L^1–scaled box function has the form

$$\tilde{\varphi}_{j,k} = 2^j \chi_{[0,1]}(2^j \cdot -k). \tag{2.5}$$

In the sequel, we will explain how to decompose the cell averages into averages of a coarser partition and details. To this end, we consider in analogy to (2.2) the two–scale relation

$$\tilde{\varphi}_{j,k} = \frac{1}{2}\left(\tilde{\varphi}_{j+1,2k} + \tilde{\varphi}_{j+1,2k+1}\right) \tag{2.6}$$

by which a coarse–scale box function is reexpressed in terms of fine–scale box functions. We now introduce the box wavelet

$$\tilde{\psi}_{j,k} := \frac{1}{2}\left(\tilde{\varphi}_{j+1,2k} - \tilde{\varphi}_{j+1,2k+1}\right). \tag{2.7}$$

Note, that the L^2–scaled counterpart coincides with the Haar wavelet, see [Haa10] and Sect. B.2.1. Then we can write any fine–scale box function by means of the box function $\tilde{\varphi}_{j,k}$ and the box wavelet $\tilde{\psi}_{j,k}$.

$$\tilde{\varphi}_{j+1,2k} = \tilde{\varphi}_{j,k} + \tilde{\psi}_{j,k}, \qquad \tilde{\varphi}_{j+1,2k+1} = \tilde{\varphi}_{j,k} - \tilde{\psi}_{j,k}. \tag{2.8}$$

These relations are motivated by the illustrations in Fig. 2.2.

In analogy to (2.6) the cell averages satisfy

$$\hat{u}_{j,k} = \frac{1}{2}\left(\hat{u}_{j+1,2k} + \hat{u}_{j+1,2k+1}\right).$$

On the other hand, (2.8) means

$$\hat{u}_{j+1,2k} = \hat{u}_{j,k} + d_{j,k}, \qquad \hat{u}_{j+1,2k+1} = \hat{u}_{j,k} - d_{j,k}$$

where the details are defined by

$$d_{j,k} := \langle u, \tilde{\psi}_{j,k}\rangle_{[0,1]} = \frac{1}{2}\left(\hat{u}_{j+1,2k} - \hat{u}_{j+1,2k+1}\right). \tag{2.9}$$

This relation shows how to reexpress fine–scale averages from coarse–scale ones and details.

In the sequel, it will be convenient to rewrite the two–scale relations in matrix–vector form. To this end, we introduce the vectors $\tilde{\Phi}_j := (\tilde{\varphi}_{j,k})_{k\in I_j}$ and $\tilde{\Psi}_j := (\tilde{\psi}_{j,k})_{k\in I_j}$. Later we will use this notation also in the sense of a collection of functions. Then (2.6) and (2.7) read

$$\tilde{\Phi}_j^T = \tilde{\Phi}_{j+1}^T\,\widetilde{\mathsf{M}}_{j,0} \quad\text{and}\quad \tilde{\Psi}_j^T = \tilde{\Phi}_{j+1}^T\,\widetilde{\mathsf{M}}_{j,1} \tag{2.10}$$

where the columns of the so–called mask matrices $\widetilde{\mathsf{M}}_{j,0}$ and $\widetilde{\mathsf{M}}_{j,1}$ contain the filter coefficients $\frac{1}{2}$, $\frac{1}{2}$ and $\frac{1}{2}$, $-\frac{1}{2}$ of $\tilde{\varphi}_{j,k}$ and $\tilde{\psi}_{j,k}$, respectively, i.e.,

$$\widetilde{\mathsf{M}}_{j,i} = \frac{1}{2} \begin{pmatrix} 1 & 0 & \dots & \dots & 0 \\ (-1)^i & 0 & & & \vdots \\ 0 & 1 & & & \\ 0 & (-1)^i & & & \\ \vdots & \ddots & \ddots & \ddots & \vdots \\ & & & 1 & 0 \\ & & & (-1)^i & 0 \\ \vdots & & & 0 & 1 \\ 0 & \dots & \dots & 0 & (-1)^i \end{pmatrix}.$$

In analogy, the two-scale relation (2.8) becomes

$$\tilde{\Phi}_{j+1}^T = \tilde{\Phi}_j^T \widetilde{\mathsf{G}}_{j,0} + \tilde{\Psi}_j^T \widetilde{\mathsf{G}}_{j,1} \qquad (2.11)$$

where the mask matrices are determined by $\widetilde{\mathsf{G}}_{j,i} = 2\,\widetilde{\mathsf{M}}_{j,i}^T$, $i = 0, 1$. Note, that the relations (2.10) and (2.11) realize a change of basis, because the composed matrices

$$\widetilde{\mathsf{M}}_j := \left(\widetilde{\mathsf{M}}_{j,0}, \widetilde{\mathsf{M}}_{j,1}\right) \quad \text{and} \quad \widetilde{\mathsf{G}}_j := \left(\widetilde{\mathsf{G}}_{j,0}^T, \widetilde{\mathsf{G}}_{j,1}^T\right)^T \qquad (2.12)$$

are inverse, i.e., $\widetilde{\mathsf{G}}_j = \widetilde{\mathsf{M}}_j^{-1}$. Hence, the single blocks fulfill

$$\widetilde{\mathsf{M}}_{j,0}\,\widetilde{\mathsf{G}}_{j,0} + \widetilde{\mathsf{M}}_{j,1}\,\widetilde{\mathsf{G}}_{j,1} = \mathsf{I}, \qquad \widetilde{\mathsf{G}}_{j,i}\,\widetilde{\mathsf{M}}_{j,i'} = \delta_{i,i'}\mathsf{I}, \quad i, i' \in \{0, 1\}. \qquad (2.13)$$

We now introduce a dual system by the functions

$$\varphi_{j,k} := 2^{-j}\,\tilde{\varphi}_{j,k} = \chi_{[0,1]}(2^j \cdot -k), \qquad \psi_{j,k} := 2^{-j}\,\tilde{\psi}_{j,k}$$

or in vector form $\Phi_j := 2^{-j}\,\tilde{\Phi}_j$ and $\Psi_j := 2^{-j}\,\tilde{\Psi}_j$. These are the L^∞–normalized counterparts of the box function and the box wavelet, respectively. Obviously, the duals also satisfy two–scale relations of the form (2.10) and (2.11) with matrices $\mathsf{M}_{j,i}$ and $\mathsf{G}_{j,i}$, $i = 0, 1$. They are related to (2.12) by

$$\mathsf{M}_{j,i} = \widetilde{\mathsf{G}}_{j,i}^T \quad \text{and} \quad \mathsf{G}_{j,i} = \widetilde{\mathsf{M}}_{j,i}^T.$$

From this we infer that $\tilde{\Phi}_j \cup \tilde{\Psi}_j$ and $\Phi_j \cup \Psi_j$ are *biorthogonal*, i.e.,

$$\begin{aligned} \langle \Phi_j, \tilde{\Phi}_j \rangle_{[0,1]} &= \langle \Psi_j, \tilde{\Psi}_j \rangle_{[0,1]} = \mathsf{I}, \\ \langle \Phi_j, \tilde{\Psi}_j \rangle_{[0,1])} &= \langle \Psi_j, \tilde{\Phi}_j \rangle_{[0,1]} = \mathbf{0} \end{aligned} \qquad (2.14)$$

where we use the notation $\langle \Theta, \Phi \rangle := (\langle \theta, \varphi \rangle)_{\theta \in \Theta, \varphi \in \Phi}$.

By means of the box function and the box wavelet as well as their L^∞–normalized counterparts we now introduce a function on the unit interval that is naturally related to the array of cell averages. To this end, we consider a function $u \in L^1([0,1])$. Then the projection of u onto piecewise constants with respect to the refinement level j is determined by

$$u_j := \sum_{k\in I_j} \langle u, \tilde{\varphi}_{j,k}\rangle_{[0,1]}\, \varphi_{j,k} =: \Phi_j^T\, \hat{\mathbf{u}}_j. \tag{2.15}$$

This function can be interpreted as an approximate solution of the conservation law on a fixed time level corresponding to the discretization level j. Equivalently, the function u_j can be represented by

$$\begin{aligned} u_j &= \sum_{k\in I_{j-1}} \langle u, \tilde{\varphi}_{j-1,k}\rangle_{[0,1]}\, \varphi_{j-1,k} + \sum_{k\in I_{j-1}} \langle u, \tilde{\psi}_{j-1,k}\rangle_{[0,1]}\, \psi_{j-1,k} \\ &=: \Phi_{j-1}^T\, \hat{\mathbf{u}}_{j-1} + \Psi_{j-1}^T\, \mathbf{d}_{j-1}, \end{aligned} \tag{2.16}$$

because the two–scale relations realize a change of basis and, in particular, the systems $\Phi_j \cup \Psi_j$ and $\tilde{\Phi}_j \cup \tilde{\Psi}_j$ are biorthogonal. This representation motivates that the details can be interpreted as the update when progressing to a higher resolution level.

We now have to explain why the representation (2.16) is preferable to (2.15) for our purposes. To this end, we verify that the details may become small when the underlying function is smooth. First of all, we conclude from (2.7) and (2.1) that

$$\langle 1, \tilde{\psi}_{j,k}\rangle_{[0,1]} = 0.$$

Since the box wavelets are L^1–normalized, i.e., $||\tilde{\psi}_{j,k}||_{L^1([0,1])} = 1$, we deduce

$$\begin{aligned} |d_{j,k}| &\le \inf_{c\in \mathsf{R}} |\langle u-c, \tilde{\psi}_{j,k}\rangle_{[0,1]}| \le \inf_{c\in \mathsf{R}} ||u-c||_{L^\infty(V_{j,k})} \\ &\le C\, 2^{-j}\, ||u'||_{L^\infty(V_{j,k})}. \end{aligned} \tag{2.17}$$

From this estimate we infer that the decay of the details is proportional to 2^{-j} provided the function u is differentiable and has a moderate derivative. Hence, the details may become small in smooth regions of u whereas it gives a significant contribution in (2.16) if the gradient of u is large. This motivates to neglect all sufficiently small details in order to compress the original data such that we control the loss of accuracy. Since the compression rates depend on the decay 2^{-j} it will be convenient to improve this decay. For this purpose not only constants have to be canceled by the wavelet $\tilde{\psi}_{j,k}$ but also all polynomials p of a fixed higher degree, i.e., $\langle p, \tilde{\psi}_{j,k}\rangle_{[0,1]} = 0$. This can be achieved by means of higher order biorthogonal systems instead of piecewise constants. In case of the unit interval we refer to [CDF92, DKU99]. By means of tensor products this can be carried over to Cartesian grids. However, for general discretizations, e.g., curvilinear grids and triangulations, we have to employ a different strategy. Here we will use a technique introduced in [CDP96] which is called the change of stable completion. In the special case it will be applied here it coincides with the lifting schemes in [Swe98]. For this purpose we will first generalize the construction of the box wavelets on an arbitrary nested grid hierarchy. The resulting wavelets will serve as a starting point for the construction of modified box wavelets in order to achieve higher order vanishing moments.

2.3 Box Wavelet

The goal is to generalize the concept of box wavelets on a hierarchy of nested grids not necessarily based on Cartesian grids. For this purpose, we proceed in two steps. First of all, we introduce the box wavelet on a Cartesian grid hierarchy using tensor products of the univariate counterparts. In a second step, we construct box wavelets on a general nested grid hierarchy such that some characteristic properties for tensor products are preserved.

2.3.1 Box Wavelet on a Cartesian Grid Hierarchy

The extension to higher dimensions is straightforward provided that the grid hierarchy is determined by Cartesian grids. To this end, we consider the unit box $\Omega = [0,1]^d$ and introduce a nested grid hierarchy by means of a uniform dyadic partition of $[0,1]^d$, i.e., $V_{j,\mathbf{k}} = \prod_{i=1}^d 2^{-j}\,[k_i, k_i+1]$, $\mathbf{k} \in I_j := \prod_{i=1}^d \{0,\dots,2^j-1\}$, see Fig. 2.1. The corresponding refinement set is characterized by $\mathcal{M}_{j,\mathbf{k}} = \{2\mathbf{k}+\mathbf{e}\,;\,\mathbf{e}\in E\}$ with $E := \prod_{i=1}^d \{0,1\}^d$. In this shift–invariant case, the multivariate box function and the box wavelets can be constructed by means of tensor products. For this purpose, we introduce the convention $\tilde{\psi}_{j,k,0} := \tilde{\varphi}_{j,k}$ and $\tilde{\psi}_{j,k,1} = \tilde{\psi}_{j,k}$ where $\tilde{\varphi}_{j,k}$ and $\tilde{\psi}_{j,k}$ denote the univariate counterparts according to Sect. 2.2. Then the multivariate box functions $\tilde{\varphi}_{j,\mathbf{k}} \equiv \tilde{\psi}_{j,\mathbf{k},\mathbf{0}}$ and box wavelets $\tilde{\psi}_{j,\mathbf{k},\mathbf{e}}$, $\mathbf{e}\in E^* := E\backslash\{\mathbf{0}\}$ are defined as products of the univariate box function $\tilde{\varphi}_{j,k}$ and the univariate box wavelet $\tilde{\psi}_{j,k}$, respectively, corresponding to the cell $V_{j,\mathbf{k}}$, i.e.,

$$\tilde{\psi}_{j,\mathbf{k},\mathbf{e}}(\mathbf{x}) := \prod_{i=1}^d \tilde{\psi}_{j,k_i,e_i}(x_i), \quad \mathbf{e}\in E. \tag{2.18}$$

These functions are shown in Fig. 2.3 where (x_1,x_2) corresponds to (x,y).

In analogy to the two–scale relations (2.6) and (2.7) the box function and the box wavelets can be rewritten in terms of fine–scale functions. To this end, we first observe that a cell $V_{j,\mathbf{k}}$ can be decomposed by $V_{j,\mathbf{k}} = \bigcup_{\mathbf{e}\in E} V_{j+1,2\mathbf{k}+\mathbf{e}}$. For any position $\mathbf{x}\in V_{j+1,2\mathbf{k}+\mathbf{i}}$ we then obtain

$$\tilde{\psi}_{j,\mathbf{k},\mathbf{e}}(\mathbf{x}) = 2^{jd}\,(-1)^{\mathbf{i}\cdot\mathbf{e}} = 2^d\,\tilde{\varphi}_{j+1,2\mathbf{k}+\mathbf{i}}(\mathbf{x})$$

where we employ (2.5) and (2.18). Finally, we conclude the two–scale relation

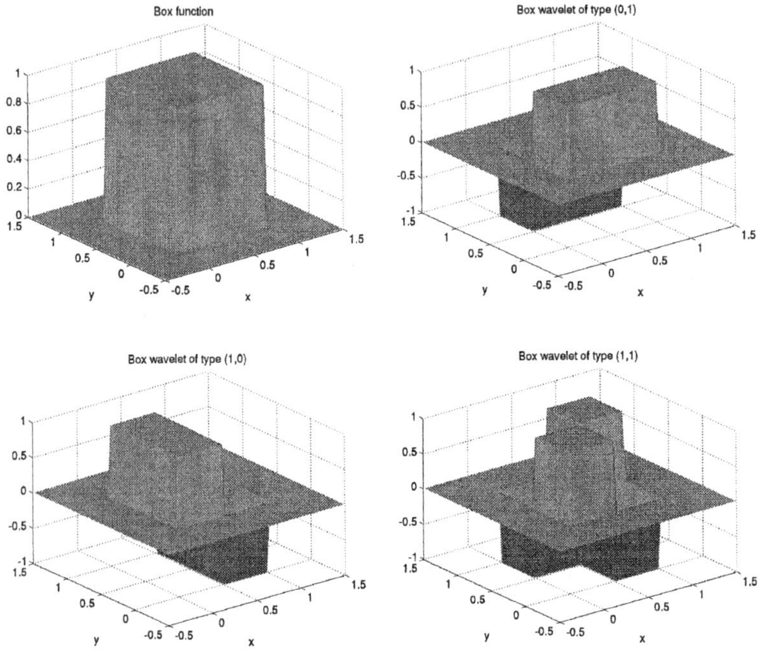

Fig. 2.3. Box function and box wavelets on $[0,1]^2$

$$\tilde{\psi}_{j,\mathbf{k},\mathbf{e}} = \sum_{\mathbf{i}\in E} 2^{-d}\,(-1)^{\mathbf{i}\cdot\mathbf{e}}\,\tilde{\varphi}_{j+1,2\mathbf{k}+\mathbf{i}}. \tag{2.19}$$

As has been proven in [Got98], p. 77, an inverse two–scale relation

$$\tilde{\varphi}_{j+1,2\mathbf{k}+\mathbf{i}} = \sum_{\mathbf{e}\in E} (-1)^{\mathbf{i}\cdot\mathbf{e}}\,\tilde{\psi}_{j,\mathbf{k},\mathbf{e}}, \quad \mathbf{i}\in E$$

holds in analogy to (2.8). Hence $\{\tilde{\varphi}_{j+1,2\mathbf{k}+\mathbf{e}}\}_{\mathbf{e}\in E}$ and $\{\tilde{\psi}_{j,\mathbf{k},\mathbf{e}}\}_{\mathbf{e}\in E}$, respectively, are systems of linearly independent functions that span the same space.

By the construction of the multivariate box functions we furthermore conclude that the multivariate box wavelets are orthogonal to constant functions, i.e.,

$$\langle 1, \tilde{\psi}_{j,\mathbf{k},\mathbf{e}}\rangle_\Omega = 0, \qquad \mathbf{e}\in E^*.$$

Finally, a biorthogonal system is determined by the L^∞–scaled counterparts

$$\psi_{j,\mathbf{k},\mathbf{e}} := 2^{-jd}\,\tilde{\psi}_{j,\mathbf{k},\mathbf{e}}, \qquad \mathbf{e}\in E,$$

according to the univariate case.

2.3.2 Box Wavelet on an Arbitrary Nested Grid Hierarchy

The goal is to generalize the concept of box wavelets on a hierarchy of nested grids not necessarily based on Cartesian grids. The construction principles are (i) the system $\tilde{\Phi}_j \cup \tilde{\Psi}_j$ is another basis for the space spanned by the basis $\tilde{\Phi}_{j+1}$, (ii) the box wavelets are orthogonal to constant functions, i.e., $\langle 1, \tilde{\Psi}_j \rangle_\Omega = \mathbf{0}$ and (iii) the basis $\tilde{\Phi}_j \cup \tilde{\Psi}_j$ and its L^∞–scaled counterpart $\Phi_j \cup \Psi_j$ are biorthogonal, i.e., (2.14) holds. For the construction we now proceed analogously to the Cartesian case as has been outlined in the previous section. To this end, we consider the box wavelet as a linear combination of the fine–scale box functions $\tilde{\varphi}_{j+1,r}$, $r \in \mathcal{M}_{j,k}$, related to the refinement of the cell $V_{j,k}$, cf. (2.19),

$$\check{\psi}_{j,k,e} := \sum_{r \in \mathcal{M}_{j,k}} \left(\frac{|V_{j+1,r}|}{|V_{j,k}|} \right)^{1/2} a_{e,r}^{j,k}\, \tilde{\varphi}_{j+1,r}, \quad e \in E^* := E \backslash \{0\} \tag{2.20}$$

with $E := \{0, \ldots, M_r - 1\}$ and appropriate parameters $a_{e,r}^{j,k}$ which have yet to be determined. Here the weights $(|V_{j+1,r}|/|V_{j,k}|)^{1/2}$ are motivated by the requirement that the box wavelets and their L^∞–scaled counterparts are biorthogonal, i.e., the identities (2.14) hold[1]. In order to distinguish these functions from their modifications to be introduced in the subsequent section we use the notation $\check{\psi}_{j,k,e}$ instead of $\tilde{\psi}_{j,k,e}$. Then we define

$$\mathbf{a}_0^{j,k} := \left(\sqrt{|V_{j+1,r}|/|V_{j,k}|} \right)_{r \in \mathcal{M}_{j,k}}, \quad \mathbf{a}_e^{j,k} := (a_{e,r}^{j,k})_{r \in \mathcal{M}_{j,k}}, \; e \in E^* \tag{2.21}$$

and the vectors of local basis functions

$$\tilde{\mathbf{\Phi}}_{j+1,k} := (\tilde{\varphi}_{j+1,r})_{r \in \mathcal{M}_{j,k}}, \qquad \check{\mathbf{\Psi}}_{j,k} := (\check{\psi}_{j,k,e})_{e \in E}.$$

Here we use the convention $\check{\psi}_{j,k,0} := \tilde{\varphi}_{j,k}$. Thus we can rewrite (2.20) in matrix–vector form as

$$\check{\mathbf{\Psi}}_{j,k} = \mathsf{A}_{j,k}\, \mathbf{\Lambda}_{j,k}\, \tilde{\mathbf{\Phi}}_{j+1,k}, \tag{2.22}$$

where the square matrices (note: $\# \mathcal{M}_{j,k} = M_r = \# E$) are defined by

$$\mathsf{A}_{j,k} := (a_{e,r}^{j,k})_{e \in E, r \in \mathcal{M}_{j,k}}, \qquad \mathbf{\Lambda}_{j,k} := \operatorname{diag}((a_{0,r}^{j,k})_{r \in \mathcal{M}_{j,k}}). \tag{2.23}$$

In order to realize a change of basis, i.e., we obtain two–scale relations of the form (2.8), we obviously need that the matrix $\mathsf{A}_{j,k}$ is invertible, i.e., the inverse denoted by

---

[1] Choosing the ansatz $\check{\psi}_{j,k,e} = \sum_{r \in \mathcal{M}_{j,k}} c_{e,r}^{j,k}\, \tilde{\varphi}_{j+1,r}$ the parameters $c_{e,r}^{j,k}$ are subject to the conditions (1) $\mathbf{c}_e^{j,k} \perp \mathbf{1}$, $e \in E^*$, i.e., $\langle \mathbf{1}, \check{\psi}_{j,k,e} \rangle_\Omega = 0$ and (2) $\langle \psi_{j,k,e}, \check{\psi}_{j,k,e'} \rangle_\Omega = \delta_{e,e'}\, c_{j,k,e}$. These conditions hold for $c_{e,r}^{j,k} = a_{0,r}^{j,k}\, a_{e,r}^{j,k}$.

$$\mathsf{A}_{j,k}^{-1} := (b_{r,e}^{j,k})_{r\in\mathcal{M}_{j,k}, e\in E}$$

exists. In this case we obtain

$$\widetilde{\mathbf{\Phi}}_{j+1,k} = \mathbf{\Lambda}_{j,k}^{-1}\,\mathsf{A}_{j,k}^{-1}\,\check{\mathbf{\Psi}}_{j,k}. \tag{2.24}$$

However, we are not only interested in performing a change of basis but the resulting system of box functions and box wavelets and their L^∞–scaled counterparts defined by

$$\varphi_{j,k} := |V_{j,k}|\,\tilde{\varphi}_{j,k}, \quad \psi_{j,k,e} := |V_{j,k}|\,\check{\psi}_{j,k,e} \tag{2.25}$$

are biorthogonal. For this purpose we prove the following corollary.

Corollary 1. *(Biorthogonality of box wavelets)*
Assume that the vectors $\mathbf{a}_e^{j,k}$, $e \in E$, form an orthonormal system, i.e., $(\mathbf{a}_e^{j,k})^T\mathbf{a}_{e'}^{j,k} = \delta_{e,e'}$. Then the systems $\check{\mathbf{\Psi}}_{j,k}$ and $\mathbf{\Psi}_{j,k} := (\psi_{j,k,e})_{e\in E}$ are biorthogonal, i.e., $\langle \mathbf{\Psi}_{j,k}, \check{\mathbf{\Psi}}_{j,k}\rangle_\Omega = \mathsf{I}$.

Proof. In order to prove the assertion we consider first

$$\langle \psi_{j,k,e}, \check{\psi}_{j,k,e'}\rangle_\Omega = |V_{j,k}| \sum_{r\in\mathcal{M}_{j,k}} \sum_{r'\in\mathcal{M}_{j,k}} a_{0,r}^{j,k}\, a_{e,r}^{j,k} a_{0,r'}^{j,k}\, a_{e',r'}^{j,k} \langle \tilde{\varphi}_{j+1,r}, \tilde{\varphi}_{j+1,r'}\rangle_\Omega$$

for any $k \in I_j$, $e, e' \in E$ and $j \in \mathbb{N}_0$. By definition of the box function (2.1) we infer for $r \in \mathcal{M}_{j,k}$ and $r' \in \mathcal{M}_{j,k'}$

$$\langle \tilde{\varphi}_{j+1,r}, \tilde{\varphi}_{j+1,r'}\rangle_\Omega = \delta_{r,r'} \frac{1}{|V_{j+1,r}|} = \delta_{r,r'}\,(a_{0,r}^{j,k})^{-2} \frac{1}{|V_{j,k}|}.$$

Note, that the refinement sets are redundancy–free. This implies

$$|\mathrm{supp}(\tilde{\psi}_{j+1,r}) \cap \mathrm{supp}(\tilde{\psi}_{j+1,r'})| = 0$$

for $r \neq r'$ according to the conditions of a nested grid hierarchy, see Definition 3. In addition, we use (2.21). Thus, we conclude by the assumption that

$$\langle \psi_{j,k,e}, \check{\psi}_{j,k,e'}\rangle_\Omega = \sum_{r\in\mathcal{M}_{j,k}} a_{e,r}^{j,k}\, a_{e',r}^{j,k} = (\mathbf{a}_e^{j,k})^T \mathbf{a}_{e'}^{j,k} = \delta_{e,e'}$$

which proves the assertion. □

Note, that the vector $\mathbf{a}_0^{j,k}$ can be extended to an orthogonal system by means of the Gram–Schmidt orthogonalization process. Finally, the orthogonal vectors have to be normalized.

According to the univariate case, we now introduce the details corresponding to the box wavelet

$$\check{d}_{j,k,e} := \langle u, \check{\psi}_{j,k,e}\rangle_\Omega, \qquad e \in E^*. \tag{2.26}$$

By the definition (2.20) of the box wavelets we then infer the two–scale relation

$$\check{d}_{j,k,e} = \sum_{r\in\mathcal{M}_{j,k}} \left(\frac{|V_{j+1,r}|}{|V_{j,k}|}\right)^{1/2} a^{j,k}_{e,r}\,\hat{u}_{j+1,r}, \quad e \in E^*. \tag{2.27}$$

On the other hand, the two–scale relation (2.4) for the cell averages can similarly be written as

$$\hat{u}_{j,k} = \sum_{r\in\mathcal{M}_{j,k}} \left(\frac{|V_{j+1,r}|}{|V_{j,k}|}\right)^{1/2} a^{j,k}_{0,r}\,\hat{u}_{j+1,r}. \tag{2.28}$$

In matrix–vector form the equations (2.27) and (2.28) then read

$$\check{\mathbf{d}}_{j,k} = \mathsf{A}_{j,k}\,\mathbf{\Lambda}_{j,k}\,\hat{\mathbf{u}}_{j+1,k}.$$

Here the vectors are defined by $\hat{\mathbf{u}}_{j+1,k} := (\hat{u}_{j+1,r})_{r\in\mathcal{M}_{j,k}}$ and $\check{\mathbf{d}}_{j,k} := (\check{d}_{j,k,e})_{e\in E}$ where we use the convention $\check{d}_{j,k,0} := \hat{u}_{j,k}$. Since the matrix $\mathsf{A}_{j,k}$ is assumed to be at least invertible in order to realize a change of basis, we can locally reexpress the fine–scale averages by the coarse–scale ones and the details, i.e.,

$$\hat{\mathbf{u}}_{j+1,k} = \mathbf{\Lambda}^{-1}_{j,k}\,\mathsf{A}^{-1}_{j,k}\,\check{\mathbf{d}}_{j,k}.$$

So far, we have only considered a local change of basis. This has been possible, because the support of the box wavelets $\check{\psi}_{j,k,e}$, $e \in E$, is completely covered by the support of the box function $\tilde{\varphi}_{j,k}$. According to the univariate case we write the global change of basis in terms of the vectors $\tilde{\mathbf{\Phi}}_j$ and $\check{\mathbf{\Psi}}_{j,e} := (\check{\psi}_{j,k,e})_{k\in I_j}$, $e \in E$. In particular, we obtain

$$\tilde{\Phi}^T_j = \tilde{\Phi}^T_{j+1}\,\check{\mathsf{M}}_{j,0}, \quad \check{\Psi}^T_{j,e} = \tilde{\Phi}^T_{j+1}\,\check{\mathsf{M}}_{j,e} \tag{2.29}$$

and

$$\tilde{\Phi}^T_{j+1} = \tilde{\Phi}^T_j\,\check{\mathsf{G}}_{j,0} + \sum_{e\in E^*} \check{\Psi}^T_{j,e}\,\check{\mathsf{G}}_{j,e}. \tag{2.30}$$

Here the entries of the mask matrices are determined by

$$\check{m}^{j,e}_{r,k} = \begin{Bmatrix} a^{j,k}_{e,r}\,a^{j,k}_{0,r} & , r \in \mathcal{M}_{j,k} \\ 0 & , \text{elsewhere} \end{Bmatrix}, \quad \check{g}^{j,e}_{k,r} = \begin{Bmatrix} b^{j,k}_{r,e}/a^{j,k}_{0,r} & , r \in \mathcal{M}_{j,k} \\ 0 & , \text{elsewhere} \end{Bmatrix} \tag{2.31}$$

for $r \in I_{j+1}$, $k \in I_j$, $e \in E$, according to (2.22) and (2.24). Note, that $b^{j,k}_{r,e} = a^{j,k}_{e,r}$ provided that the matrix $\mathsf{A}_{j,k}$ is orthogonal. The sparsity pattern of the mask matrices is only presented for the two–dimensional case

$$\check{\mathsf{G}}_{j,e} = \begin{pmatrix} * * * * & & & & \\ & * * * * & & & \\ & & \ddots & & \\ & & & * * * * & \\ & & & & * * * * \end{pmatrix}$$

where we assume that each cell is decomposed into four subcells, see Fig. 2.1. The pattern of $\check{\mathsf{M}}_{j,e}$ coincides with the pattern of the transpose of $\check{\mathsf{G}}_{j,e}$. In particular, there is exactly one entry in each row and column due to the nestedness of the grid hierarchy.

We now gather all mask matrices corresponding to the box wavelets of type $e \in E^*$ in one single block, i.e.,

$$\check{\mathsf{M}}_{j,1} := \left(\check{\mathsf{M}}_{j,e}\right)_{e\in E^*} \quad \text{and} \quad \check{\mathsf{G}}_{j,1} := \left((\check{\mathsf{G}}_{j,e}^T)_{e\in E^*}\right)^T .$$

Then we can define the matrices

$$\check{\mathsf{M}}_j := \left(\check{\mathsf{M}}_{j,0}, \check{\mathsf{M}}_{j,1}\right) \quad \text{and} \quad \check{\mathsf{G}}_j := \left(\check{\mathsf{G}}_{j,0}^T, \check{\mathsf{G}}_{j,1}^T\right)^T = \check{\mathsf{M}}_j^{-1}$$

as in the univariate case, see (2.12). Note, that the change of basis (2.29) and (2.30) implies that $\check{\mathsf{M}}_j$ and $\check{\mathsf{G}}_j$ are inverse. Similar two–scale relations hold for the L^∞–normalized counterparts defined by (2.25) for the vectors

$$\boldsymbol{\Phi}_j := \mathsf{V}_j\, \tilde{\boldsymbol{\Phi}}_j \quad \text{and} \quad \boldsymbol{\Psi}_{j,e} := \mathsf{V}_j\, \check{\boldsymbol{\Psi}}_{j,e} \tag{2.32}$$

with the diagonal matrix $\mathsf{V}_j := \operatorname{diag}\left((|V_{j,k}|)_{k\in I_j}\right)$. Then we infer from the local biorthogonality property, see Corollary (1), that the systems $\tilde{\boldsymbol{\Phi}}_j \cup \bigcup_{e\in E^*} \check{\boldsymbol{\Psi}}_{j,e}$ and $\boldsymbol{\Phi}_j \cup \bigcup_{e\in E^*} \boldsymbol{\Psi}_{j,e}$ are biorthogonal provided that the matrices $\mathsf{A}_{j,k}$ are orthogonal. Hence, the projection of any function $u \in L^1(\Omega)$ onto piecewise constants with respect to the refinement level j can be represented by

$$u_j := \Phi_j^T \langle u, \tilde{\Phi}_j\rangle_\Omega = \Phi_{j-1}^T \langle u, \tilde{\Phi}_{j-1}\rangle_\Omega + \sum_{e\in E^*} \Psi_{j-1,e}^T \langle u, \tilde{\Psi}_{j-1,e}\rangle_\Omega .$$

Finally, we have to verify that the box wavelets have one vanishing moment, i.e., they are orthogonal to constant functions. To this end we note that

$$\langle 1, \check{\psi}_{j,k,e}\rangle_\Omega = \sum_{r\in\mathcal{M}_{j,k}} a_{0,r}^{j,k}\, a_{e,r}^{j,k}\, \langle 1, \tilde{\varphi}_{j+1,r}\rangle_\Omega = (\mathbf{a}_0^{j,k})^T \mathbf{a}_e^{j,k}$$

holds for all $e \in E^*$, $k \in I_j$, $j \in \mathbb{N}_0$. Again, we need that the vectors $\mathbf{a}_e^{j,k}$, $e \in E$, are at least orthogonal. According to the univariate case we then infer that the details decay like $|V_{j,k}|$ which becomes smaller with increasing refinement level j.

2.4 Change of Stable Completion

In order to improve the compression rates we need wavelets with better cancellation properties in the sense that higher order polynomial moments vanish. In [CDP96] a systematic ansatz has been proposed for this task. To this end, we first note that the construction of the matrix $\check{\mathsf{M}}_{j,1}$ is only *one* way to complement the matrix $\check{\mathsf{M}}_{j,0}$ to an invertible matrix $\check{\mathsf{M}}_j$. There is a continuum of *completions of* $\check{\mathsf{M}}_{j,0}$. We call a completion *(uniformly) stable* if the matrix $\check{\mathsf{M}}_j$ and its inverse $\check{\mathsf{G}}_j$ have uniformly bounded operator norms with respect to a suitable vector norm. For the matrices $\check{\mathsf{M}}_j$ and $\check{\mathsf{G}}_j$ corresponding to the completion by means of the box wavelet it can be proven that they are stable in any l^p–norm, see Sect. B.1.2 for $p = 2$. Note, that any completion $\widetilde{\mathsf{M}}_{j,1}$ of $\check{\mathsf{M}}_{j,0}$ characterizes a particular complement of the spaces of piecewise constants related to the refinement levels j and $j+1$. In particular, the corresponding wavelet basis $\tilde{\Psi}_j$ is stable in the sense that

$$c\,\|\mathbf{d}_j\|_{l^1} \leq \|\tilde{\mathbf{\Psi}}_j^T\,\mathbf{d}_j\|_{L^1(\Omega)} \leq C\,\|\mathbf{d}_j\|_{l^1}$$

holds for constants c, C independent of j.

In the sequel, we will construct another stable completion $\widetilde{\mathsf{M}}_{j,1}$ of $\check{\mathsf{M}}_{j,0}$ such that the corresponding wavelets have higher order vanishing moments. The starting point is any composed matrix

$$\mathsf{L}_j = (\mathsf{L}_{j,e})_{e\in E^*}$$

where the matrices $\mathsf{L}_{j,e} \in \mathsf{R}^{N_j\times N_j}$, $N_j := \#\, I_j$, are uniformly bounded with respect to l^1. Then we note that the matrices

$$\widetilde{\mathsf{M}}_j := \check{\mathsf{M}}_j \begin{pmatrix} \mathsf{I} & \mathsf{L}_j \\ 0 & \mathsf{I} \end{pmatrix} \quad \text{and} \quad \widetilde{\mathsf{G}}_j := \begin{pmatrix} \mathsf{I} & -\mathsf{L}_j \\ 0 & \mathsf{I} \end{pmatrix} \check{\mathsf{G}}_j$$

are inverse, i.e., $\widetilde{\mathsf{G}}_j = \widetilde{\mathsf{M}}_j^{-1}$. From the composed matrices we determine the single blocks

$$\widetilde{\mathsf{M}}_{j,0} = \check{\mathsf{M}}_{j,0},\ \widetilde{\mathsf{M}}_{j,1} = \check{\mathsf{M}}_{j,1} + \check{\mathsf{M}}_{j,0}\,\mathsf{L}_j,\ \widetilde{\mathsf{G}}_{j,0} = \check{\mathsf{G}}_{j,0} - \mathsf{L}_j\,\check{\mathsf{G}}_{j,1},\ \widetilde{\mathsf{G}}_{j,1} = \check{\mathsf{G}}_{j,1}$$

and, equivalently,

$$\begin{aligned} \widetilde{\mathsf{M}}_{j,0} &= \check{\mathsf{M}}_{j,0}, & \widetilde{\mathsf{M}}_{j,e} &= \check{\mathsf{M}}_{j,e} + \check{\mathsf{M}}_{j,0}\,\mathsf{L}_{j,e},\ e \in E^*, \\ \widetilde{\mathsf{G}}_{j,0} &= \check{\mathsf{G}}_{j,0} - \sum_{e\in E^*} \mathsf{L}_{j,e}\,\check{\mathsf{G}}_{j,e}, & \widetilde{\mathsf{G}}_{j,e} &= \check{\mathsf{G}}_{j,e},\ e \in E^*. \end{aligned} \tag{2.33}$$

where we employ the block structure of L_j, $\check{\mathsf{M}}_{j,1}$ and $\check{\mathsf{G}}_{j,1}$ with respect to the different wavelet types $e \in E^*$. Since the matrices $\mathsf{L}_{j,e}$, $e \in E^*$, are assumed to be uniformly bounded, the matrix $\widetilde{\mathsf{M}}_{j,1}$ is still a stable completion of $\check{\mathsf{M}}_{j,0}$. In terms of the new completion, the basis of modified box wavelets can now be represented in terms of the old wavelet basis $\check{\Psi}_j$, i.e.,

$$\tilde{\Psi}_j^T := \tilde{\Phi}_{j+1}^T \widetilde{\mathsf{M}}_{j,1} = \tilde{\Phi}_{j+1}^T \check{\mathsf{M}}_{j,1} + \tilde{\Phi}_{j+1}^T \check{\mathsf{M}}_{j,0}\mathsf{L}_j = \check{\Psi}_j^T + \tilde{\Phi}_j^T \mathsf{L}_j. \tag{2.34}$$

Note, that the basis $\tilde{\Phi}_j$ of box functions is not changed, since $\widetilde{\mathsf{M}}_{j,0} = \check{\mathsf{M}}_{j,0}$. Only the wavelet basis $\check{\Psi}_j$ is modified according to a linear combination of box functions on level j and an appropriate weighting.

Finally, we emphasize that the change of stable completion offers some free parameters for the construction of modified box wavelets with better cancellation properties. However, it is not obvious that there is a biorthogonal system $\{\Phi_j \cup \Psi_j\}_{j \in \mathbb{N}_0}$, i.e., whether there actually exist function vectors Φ_j in L^∞ satisfying two-scale relations of the type (2.29) for the $\check{\mathsf{M}}_{j,0}$ replaced by $\mathsf{M}_{j,0}$. In general, the L^∞–normalized counterparts according to (2.32) do not form an biorthogonal system. The existence of a biorthogonal basis is strongly related to the stability of the multiscale bases $\{\tilde{\Phi}_0 \cup \tilde{\Psi}_0 \cup \cdots \cup \tilde{\Psi}_{L-1}\}_{L \in \mathbb{N}_0}$ instead of the two-scale bases $\{\tilde{\Phi}_j \cup \tilde{\Psi}_j\}_{j \in \mathbb{N}_0}$. Note, that the stability of the multiscale bases can only be expected to hold in the Hilbert space L^2 while the relevant topology here is ultimately determined by L^1. For details we refer to [Dah94].

2.5 Box Wavelet with Higher Vanishing Moments

The change of stable completion according to Sect. 2.4 is now applied to the generalized box wavelets presented in Sect. 2.3. For this purpose, we first introduce the definition of higher order vanishing moments and outline an algorithm for determining the modified box wavelets. A univariate example is presented as an illustration. We conclude this section with a result on compression rates.

2.5.1 Definition and Construction

We will now determine the free parameters introduced by the matrices $\mathsf{L}_{j,e}$, $e \in E^*$, such that the modified box wavelets have higher order vanishing moments. To this end, we introduce the following notion.

Definition 4. *(Vanishing Moments)*
The wavelet basis has M vanishing moments *if*

$$\langle p, \tilde{\psi}_{j,k,e} \rangle_\Omega = 0$$

holds for all polynomials $p \in \mathsf{P}_{M-1}$ of degree less than M and $e \in E^*$, $k \in I_j$, $j \in \mathbb{N}_0$.

In order to ensure an efficient two–scale transformation between the bases of box functions $\tilde{\Phi}_{j+1}$ and the two–scale basis $\tilde{\Phi}_j \cup \tilde{\Psi}_j$ it will be convenient to choose only a small number of parameters, i.e., the number of non–vanishing entries in each column and row of the matrices $\mathsf{L}_{j,e}$, $e \in E^*$, should be uniformly bounded. This is related to a local support of the modified box wavelets which is uniformly bounded.

The construction principle follows the idea presented first in [Got98], pp. 78. It is summarized in the following algorithm.

Algorithm 1. *(Realization of higher order vanishing moments)*

1. *Choose the order of vanishing moments M and determine the number of required conditions $M_c := \binom{M+d-1}{d} = \dim \mathsf{P}_{M-1}$;*
2. *choose the stencil $\mathcal{L}^e_{j,k} \subset I_j$, of box functions by which the box wavelet is modified, i.e.,*
$$\tilde{\psi}_{j,k,e} = \check{\psi}_{j,k,e} + \sum_{l \in \mathcal{L}^e_{j,k}} l^{j,e}_{l,k}\, \tilde{\varphi}_{j,l},\ e \in E^*,$$
such that $M_c \leq \# \mathcal{L}^e_{j,k} \leq M_{\mathcal{L}}$ and $M_{\mathcal{L}}$ independent of j, k and e;
3. *choose a basis $\{\omega_i\}_{i\in\mathcal{P}_{M-1}}$ for P_{M-1} with $\mathcal{P}_{M-1} := \{1,\dots,M_c\}$;*
4. *determine the free parameters $l^{j,e}_{l,k}$, $l \in \mathcal{L}^e_{j,k}$, such that*
$$\langle \omega_i, \tilde{\psi}_{j,k,e}\rangle_\Omega = 0 \quad \forall\, i \in \mathcal{P}_{M-1}. \tag{2.35}$$

The system (2.35) results in a linear system of equations $\mathsf{A}\,\mathbf{l} = \mathbf{b}$ for the unknowns $\mathbf{l} = (l^{j,e}_{l,k})_{l\in\mathcal{L}^e_{j,k}}$. Here we omit the dependence on j, k, e. The matrix A and the right hand side $\mathbf{b}$ are determined by

$$\mathsf{A} = \big(\langle \omega_i, \tilde{\varphi}_{j,l}\rangle_\Omega\big)_{i\in\mathcal{P}_{M-1},\, l\in\mathcal{L}^e_{j,k}}, \qquad \mathbf{b} = \big(-\langle \omega_i, \check{\psi}_{j,k,e}\rangle_\Omega\big)_{i\in\mathcal{P}_{M-1}}.$$

Introducing the moments with respect to the basis $\{\omega_i\}_{i\in\mathcal{P}_{M-1}}$, i.e.,

$$\nu_{j,l,i} := \langle \omega_i, \chi_{V_{j,l}}\rangle_\Omega,$$

these inner products can be rewritten as

$$\langle \omega_i, \tilde{\varphi}_{j,l}\rangle_\Omega = \frac{1}{|V_{j,l}|}\nu_{j,l,i}, \quad \langle \omega_i, \check{\psi}_{j,k,e}\rangle_\Omega = \sum_{r\in\mathcal{M}_{j,k}} \frac{a^{j,k}_{0,r}\, a^{j,k}_{e,r}}{|V_{j+1,r}|}\, \nu_{j+1,r,i}.$$

Then the linear system (2.35) can be reexpressed as

$$\sum_{l\in\mathcal{L}^e_{j,k}} \frac{|V_{j,k}|}{|V_{j,l}|}\, \nu_{j,l,i}\, l^{j,e}_{l,k} = -\sum_{r\in\mathcal{M}_{j,k}} \frac{a^{j,k}_{e,r}}{a^{j,k}_{0,r}}\, \nu_{j+1,r,i}.$$

Concerning the choice of an appropriate basis for P_{M-1} we might choose the monomials

$$\omega_{\mathbf{i}}(\mathbf{x}) = \mathbf{x}^{\mathbf{i}} := \prod_{l=1}^{d} x_l^{i_l}.$$

However, it might be more useful to apply a local set of basis functions, e.g.,

$$\omega_{\mathbf{i}}(\mathbf{x}) = (\mathbf{x} - \mathbf{c}_{j,k})^{\mathbf{i}}, \qquad \mathbf{c}_{j,k} = \frac{1}{|V_{j,k}|}\int_{V_{j,k}} \mathbf{x}\, d\mathbf{x}, \tag{2.36}$$

where $\mathbf{c}_{j,k}$ denotes the center of mass and $\mathbf{i} \in N_0^d$, $0 \le \sum_{k=1}^d i_k \le M-1$.

Since the solution of the linear system is not uniquely determined, the underdetermined system (2.35) is solved by means of singular value decomposition, cf. [GV96]. In practice, this solver proves to be very reliable but expensive.

According to (2.33) the entries of the mask matrices related to the modified box wavelets are determined by

$$\tilde{m}_{r,k}^{j,0} = \check{m}_{r,k}^{j,0}, \quad \tilde{m}_{r,k}^{j,e} = \check{m}_{r,k}^{j,e} + l_{l,k}^{j,e}\,\check{m}_{r,l}^{j,0}, \quad e \in E^*,$$
$$\tilde{g}_{k,r}^{j,0} = \check{g}_{k,r}^{j,0} - \sum\nolimits_{e\in E^*} l_{k,l}^{j,e}\,\check{g}_{l,r}^{j,e}, \quad \tilde{g}_{k,r}^{j,e} = \check{g}_{k,r}^{j,e}, \quad e \in E^*, \tag{2.37}$$

for $r \in I_{j+1}$, $k \in I_j$. Here the index $l \in I_j$ is uniquely determined by $r \in \mathcal{M}_{j,l}$. Moreover we use the convention that $l_{l,k}^{j,e} = 0$ if $l \notin \mathcal{L}_{j,k}^e$. In more detail, the modified coefficients read

$$\tilde{m}_{r,k}^{j,e} = \begin{cases} (a_{0,r}^{j,l})^2 \left(\delta_{l,k}\,(a_{0,r}^{j,l})^{-1}\, a_{e,r}^{j,l} + l_{l,k}^{j,e}\right) & , r \in \mathcal{M}_{j,l},\ l \in \mathcal{L}_{j,k}^e \cup \{k\}, \\ 0 & , \text{elsewhere}, \end{cases}$$
$$\tilde{g}_{k,r}^{j,0} = \begin{cases} (a_{0,r}^{j,l})^{-1} \left(\delta_{l,k}\, b_{r,0}^{j,l} - \sum\nolimits_{e\in E^*} b_{r,e}^{j,l}\, l_{k,l}^{j,e}\right) & , r \in \mathcal{M}_{j,l},\ l \in I_j, \\ & \quad k \in \mathcal{L}_{j,l}^e \cup \{l\}, \\ 0 & , \text{elsewhere}, \end{cases} \tag{2.38}$$

for $r \in I_{j+1}$, $k \in I_j$, $e \in E^*$. .

2.5.2 A Univariate Example

We now apply Algorithm 1 to the univariate box wavelet $\check{\psi}_{j,k}$ on the unit interval, cf. Sect. 2.2, where we consider a uniform dyadic partition of $\Omega = [0,1]$, i.e., $V_{j,k} = 2^{-j}[k,k+1]$, $k \in I_j = \{0,\dots,2^j-1\}$ and $\mathcal{M}_{j,k} = \{2k, 2k+1\}$. For simplicity, we now assume that the order of vanishing moments is odd, i.e., $M = 2s+1 = M_c$ for some $s \in \mathsf{N}_0$. In particular, the dimension of P_{M-1} coincides with the order of vanishing moments, i.e., $M = M_c$, that only holds in the one–dimensional case. Since there is only one wavelet type, we will omit the index e for the moment. The univariate modified box wavelet $\tilde{\psi}_{j,k}$ is now determined by the stencil

$$\mathcal{L}_{j,k} = \{k-s+\mu, \dots, k+s+\mu\} \tag{2.39}$$

of box functions $\tilde{\varphi}_{j,l}$ that are added to the box wavelet $\check{\psi}_{j,k}$ weighted by the still unknown parameters $l_{l,k}$, $l \in \mathcal{L}_{j,k}$. Note that the stencil $\mathcal{L}_{j,k}$ is chosen to be symmetric ($\mu = 0$) for $k \in \{k-s, \dots, 2^j-1-s\}$ but has to be one–sided near the interval boundaries such that the support of $\tilde{\psi}_{j,k}$ is still

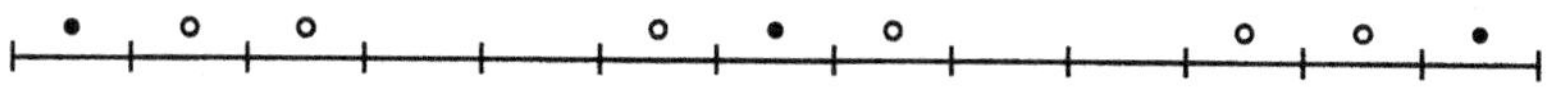

Fig. 2.4. Illustration of the sets $\mathcal{L}_{j,k}$ for $s = 1$

contained in $[0,1]$, i.e., $\mu \in \{-s, \dots, s\}$. For $s = 1$ the three different stencils are illustrated in Fig. 2.4. Here • indicates the support of the box wavelet $\check{\psi}_{j,k}$ and ∘ indicates the extended support of the modified box wavelet $\tilde{\psi}_{j,k}$. In particular, the stencils from the left to the right correspond to $\mu = 1, 0, -1$, respectively. We now determine the coefficients $l_{l,k}$, $l \in \mathcal{L}_{j,k}$. For this purpose, the parameters $s \in \mathsf{N}_0$ and $\mu \in \{-s, \dots, s\}$ are arbitrarily but fixed. Note that j has to be chosen such that $2^j \geq M = 2s + 1$ otherwise the support of the modified box wavelet $\tilde{\psi}_{j,k}$ is not contained in $[0,1]$. According to (2.36) we put $\omega_i(x) := (x - 2^{-j}(k + 0.5))^i$. Then the system (2.35) reads

$$\sum_{l=-s+\mu}^{s+\mu} 2^{i+1}\left((l+0.5)^{i+1} - (l-0.5)^{i+1}\right) l_{l+k,k} = 1 - (-1)^i$$

for $i = 0, \dots, M - 1 = 2s$. Obviously, the solution depends only on the position k and the parameters s and μ, respectively, but is independent of the level j due to the uniform dyadic grid hierarchy. In particular, the system of linear equations is regular. For $s = 1$ and $s = 2$ the solution is summarized in Tables 2.1 and 2.2. The corresponding mask matrix $\mathsf{L} \equiv \mathsf{L}_{j,1}$ for the case $s = 1$ is determined by the coefficients in Table 2.1 and the stencil (2.39) and reads

Table 2.1. Coefficients for $s = 1$

$\mu = -1$	$\mu = 0$	$\mu = 1$
$\frac{1}{8}$	$-\frac{1}{8}$	$-\frac{3}{8}$
$-\frac{1}{2}$	0	$\frac{1}{2}$
$\frac{3}{8}$	$\frac{1}{8}$	$-\frac{1}{8}$

Table 2.2. Coefficients for $s = 2$

$\mu = -2$	$\mu = -1$	$\mu = 0$	$\mu = 1$	$\mu = 2$
$\frac{7}{128}$	$-\frac{3}{128}$	$\frac{3}{128}$	$-\frac{7}{128}$	$-\frac{65}{128}$
$-\frac{19}{64}$	$\frac{9}{64}$	$-\frac{11}{64}$	$-\frac{15}{64}$	$\frac{61}{64}$
$\frac{11}{16}$	$-\frac{13}{32}$	0	$\frac{13}{32}$	$-\frac{11}{16}$
$-\frac{61}{64}$	$\frac{15}{64}$	$\frac{11}{64}$	$-\frac{9}{64}$	$\frac{19}{64}$
$\frac{65}{128}$	$\frac{7}{128}$	$-\frac{3}{128}$	$\frac{3}{128}$	$-\frac{7}{128}$

$$\mathsf{L}_{j,e} = \frac{1}{8} \begin{pmatrix} -3 & -1 & & & & & \\ 4 & 0 & -1 & & & & \\ -1 & 1 & 0 & & & & \\ & & 1 & \ddots & & & \\ & & & \ddots & & & \\ & & & \ddots & & & \\ & & & & -1 & & \\ & & & & 0 & -1 & 1 \\ & & & & 1 & 0 & -4 \\ & & & & & 1 & 3 \end{pmatrix}.$$

Finally, in Fig. 2.5 we present the box wavelet $\check{\psi}_{j,k}$ and the modified box wavelets $\tilde{\psi}_{j,k}$ for $s = 1$ and $\mu = -1, 0, 1$.

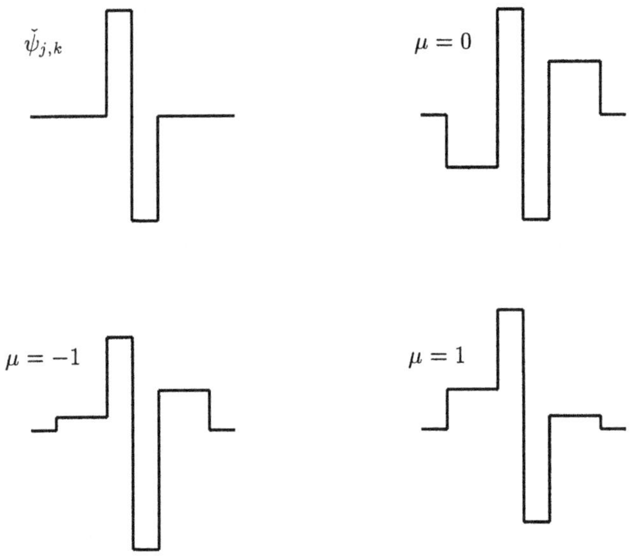

Fig. 2.5. Box wavelet $\check{\psi}_{j,k}$ and modified box wavelets $\tilde{\psi}_{j,k}$ for $s = 1$ and $\mu = -1, 0, 1$

2.5.3 A Remark on Compression Rates

Since the construction of modified box wavelets is motivated by higher compression rates, we present a standard result for the decay of the details. It can be proven similarly to (2.17) where we employ regularity of the function u and the cancellation property of the modified box wavelets.

Corollary 2. *(Regularity $\Rightarrow$ Compression)*
Assume that the support of the modified box wavelets $\tilde{\psi}_{j,k,e}$ is uniformly bounded, i.e., $\kappa_{j,k,e} := \operatorname{supp} \tilde{\psi}_{j,k,e}$ and $\operatorname{diam} \kappa_{j,k,e} \sim 2^{-j}$, and the modified box wavelets are L^p–normalized, i.e., $\|\tilde{\psi}_{j,k,e}\|_{L^p(\Omega)} \sim 1$ where $1 \leq p \leq \infty$. For a function $u \in W^{q,M}(\kappa_{j,k})$, $1/p + 1/q = 1$, the wavelet coefficients can be estimated by

$$|d_{j,k,e}| = |\langle u, \tilde{\psi}_{j,k,e}\rangle_\Omega| \leq c\, 2^{-jM} \|u\|_{W^{q,M}(\kappa_{j,k,e})}$$

provided that the wavelet $\tilde{\psi}_{j,k,e}$ has M vanishing moments. Here $c > 0$ is a constant independent of $j \in \mathbb{N}_0$, $k \in I_j$, $e \in E^*$ and u.

For our purposes we need $p = 1$ and $q = \infty$. Obviously, the size of $|d_{j,k,e}|$ decreases with increasing level where the decay is the stronger the higher the number of vanishing moments is. However, the higher the number of vanishing moments the more parameters are needed for their realization, i.e., $\# \mathcal{L}^e_{j,k}$ increases. Therefore the realization of the change of basis is more expensive. On the other hand, the compression rates may be higher. Therefore the number of vanishing moments should be not chosen too large.

Finally, we emphasize that the resulting completion $\widetilde{\mathsf{M}}_{j,1}$ of the initial stable completion $\check{\mathsf{M}}_{j,0}$ is stable in the sense specified in Sect. 2.4. However, it has not been verified so far that there exists a biorthogonal system for an arbitrary hierarchy of nested grids.

2.6 Multiscale Transformation

As already outlined in Section 2.2 the ultimate goal is to transform an array of cell averages into a different format in order to compress data. This is achieved by means of a change of basis. For this purpose we introduce the vectors

$$\hat{\mathbf{u}}_j := (\hat{u}_{j,k})_{k \in I_j} \quad \text{and} \quad \mathbf{d}_j := (d_{j,k,e})_{k \in I_j, e \in E^*}.$$

Here the averages and the details are determined by functionals of a function $u \in L^1(\Omega)$, i.e., $\hat{u}_{j,k} := \langle u, \tilde{\varphi}_{j,k}\rangle_\Omega$ and $d_{j,k,e} := \langle u, \tilde{\psi}_{j,k,e}\rangle_\Omega$. According to (2.29), (2.30) and (2.34) we obtain the two–scale relations

$$\hat{\mathbf{u}}_{j+1} = \widetilde{\mathsf{G}}_j^T \begin{pmatrix} \hat{\mathbf{u}}_j \\ \mathbf{d}_j \end{pmatrix}, \qquad \begin{pmatrix} \hat{\mathbf{u}}_j \\ \mathbf{d}_j \end{pmatrix} = \widetilde{\mathsf{M}}_j^T \hat{\mathbf{u}}_{j+1}. \tag{2.40}$$

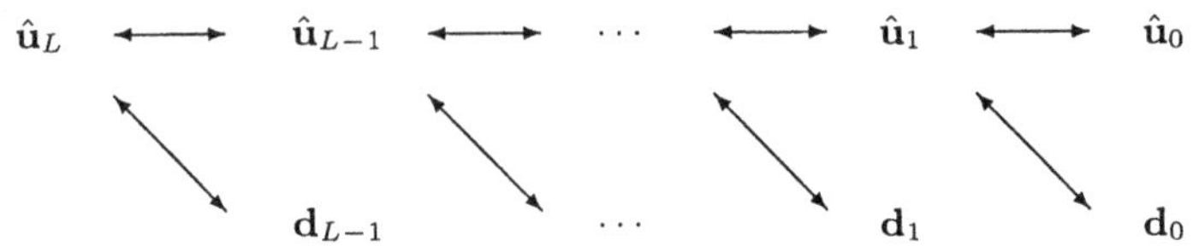

Fig. 2.6. Pyramid scheme of multiscale transformation

Applying these relations iteratively, see Fig. 2.6, the array $\hat{\mathbf{u}}_L$ of cell averages on level L can be decomposed into a sequence of coarse–scale cell averages $\hat{\mathbf{u}}_0$ and details $\mathbf{d}_j$, $j = 0, \ldots, L-1$. It will be convenient to introduce the notation

$$\mathbf{d}^{(L)} := (\mathbf{d}_{-1}^T, \mathbf{d}_0^T, \ldots, \mathbf{d}_{L-1}^T)^T, \qquad \mathbf{d}_{-1} := \hat{\mathbf{u}}_0 \tag{2.41}$$

for the multiscale coefficients. Since the bases $\{\tilde{\Phi}_j\}_{j\in\mathbb{N}_0}$ and $\{\tilde{\Phi}_j \cup \tilde{\Psi}_j\}_{j\in\mathbb{N}_0}$ are assumed to be uniformly stable in the sense of Section 2.4, there is a bijective linear operator $\mathcal{M}_L : \mathbb{R}^{N_L} \to \mathbb{R}^{N_L}$ by which the transformation between the single–scale coefficients $\hat{\mathbf{u}}_L$ and the multiscale coefficients $\mathbf{d}^{(L)}$ can be carried out, i.e.,

$$\mathbf{d}^{(L)} = \mathcal{M}_L \hat{\mathbf{u}}_L \quad \text{and} \quad \hat{\mathbf{u}}_L = \mathcal{M}_L^{-1} \mathbf{d}^{(L)}.$$

The operator $\mathcal{M}_L$ is called the *multiscale transformation* and $\mathcal{M}_L^{-1}$ the *inverse multiscale transformation*. From the two–scale relations (2.40) we deduce the following matrix representation of the multiscale transformations

$$\mathcal{M}_L = \overline{\mathsf{M}}_0 \cdots \overline{\mathsf{M}}_{L-1}, \qquad \mathcal{M}_L^{-1} = \overline{\mathsf{G}}_{L-1} \cdots \overline{\mathsf{G}}_0 \tag{2.42}$$

with the matrices

$$\overline{\mathsf{M}}_j := \begin{pmatrix} \widetilde{\mathsf{M}}_j^T & \mathbf{0} \\ \mathbf{0} & \mathsf{I}_j \end{pmatrix}, \qquad \overline{\mathsf{G}}_j := \begin{pmatrix} \widetilde{\mathsf{G}}_j^T & \mathbf{0} \\ \mathbf{0} & \mathsf{I}_j \end{pmatrix}, \qquad \mathsf{I}_j := \mathsf{I}_{N_L - N_{j+1}}.$$

Provided that there is an biorthogonal system

$$\mathbf{\Psi}^{(L)} := \Phi_0 \cup \bigcup_{j=0}^{L-1} \Psi_j$$

to the multiscale basis

$$\tilde{\mathbf{\Psi}}^{(L)} := \tilde{\Phi}_0 \cup \bigcup_{j=0}^{L-1} \tilde{\Psi}_j,$$

then the multiscale transformation realizes the change of basis, i.e.,

$$u_L := \Phi_L^T \, \hat{\mathbf{u}}_L = \Phi_0^T \, \hat{\mathbf{u}}_0 + \sum_{j=0}^{L-1} \Psi_j^T \, \mathbf{d}_j . \tag{2.43}$$

Here u_L is the projection of any function $u \in L^1(\Omega)$ onto the space spanned by the basis Φ_L. As already discussed in Section 2.4 the existence of the biorthogonal system is related to the uniform stability of the L^2normalized counterparts of the modified box wavelets, see [Dah94].

In the sequel, the basis functions Φ_j and Ψ_j are always referred to as *primal scaling functions* and *primal wavelets*, respectively. Analogously, the basis functions $\tilde{\Phi}_j$ and $\tilde{\Psi}_j$ are called the *dual scaling functions* and *dual wavelet*, respectively. This terminology is often used to express that the functions under consideration are expanded in the primal bases while bounded linear functionals are represented in the dual basis. In the present context we are interested in multiscale transformations of arrays of cell averages, i.e., of *functionals* of the unknown solution. This motivates our notation for the box functions and box wavelets introduced above. Accordingly we will actually explicitly construct always only the uniformly stable bases $\{\tilde{\Phi}_j\}_{j\in\mathbb{N}_0}$ and $\{\tilde{\Psi}_j\}_{j\in\mathbb{N}_0}$ that are dual in the above sense to process the cell averages. The corresponding primal bases will enter only the analysis but will never be needed explicitly for algorithmic realizations.

Finally, we would like to conclude with some remarks concerning the efficiency of the multiscale transformation. In order to keep the transformation efficient it is not reasonable to compute the matrix $\mathcal{M}_L$ given by (2.42). Instead the transformation should be performed by applying successively the two–scale transformation (2.40) as illustrated in Fig. 2.6. The best we can hope for is an effort proportional to the dimension of the finest space, i.e., $\mathcal{O}(N_L)$. For this purpose, the mask matrices $\widetilde{\mathsf{M}}_j$ and $\widetilde{\mathsf{G}}_j$ should be *uniformly banded*, i.e., in each row and column there are only a uniformly bounded number of non–vanishing entries. This task is closely connected to the construction of basis functions which are *locally finite*, i.e., the number of functions of level j that do not vanish in $\mathbf{x} \in \Omega$ is uniformly bounded. If we now assume that the dimension of the spaces increases at least by a constant factor, i.e., $N_{j+1} \geq a\, N_j$, $a > 1$, then the multiscale transformation can be carried out in $\mathcal{O}(N_L)$ operations. Note, that it is still prohibited to compute $\mathcal{M}_L$ because products of sparse matrices are in general not sparse.

3 Locally Refined Spaces

In general, a uniform refinement of the discretization is not adequate, since this results in a huge number of discretization points also in regions where the solution is smooth and a coarser grid would be sufficient for appropriately resolving the solution. Instead, the grid should be adapted to the problem at hand to produce an economic representation of the solution. By means of local grid refinements the complexity of the problem, i.e., the number of discretization points or unknowns, can be reduced to the number of "significant" points. The meaning of significance will be clarified later. We will see that locally adapted grids correspond in a natural way to locally refined spaces spanned by box functions or wavelet functions, i.e., $\tilde{\Phi}_{L,\varepsilon} \subset \bigcup_{j=0}^{L} \tilde{\Phi}_j$ and $\tilde{\Psi}_{L,\varepsilon} \subset \bigcup_{j=-1}^{L-1} \tilde{\Psi}_j$, respectively. Here the refinement strategy will be based on the multiscale decomposition corresponding to the modified box wavelets. By means of a truncated sequence of details corresponding to the index set $\mathcal{D}_{L,\varepsilon} \subset \bigcup_{j=0}^{L-1} J_j$, $J_j := \{j\} \times I_j \times E^*$, we will determine an adapted grid on which the evolution step will be performed. This is outlined in Sect. 3.1. To facilitate an efficient change of bases between the local spaces $\tilde{\Phi}_{L,\varepsilon}$ and $\tilde{\Psi}_{L,\varepsilon}$ the notion of *grading* is introduced. We will investigate in Sect. 3.2 how the grading of $\mathcal{D}_{L,\varepsilon}$ affects the grading of the adapted grid. The local multiscale transformation and its inverse are presented in Sect. 3.3. We emphasize that the number of operations is proportional to $\#\mathcal{D}_{L,\varepsilon}$ instead of N_L. The feasibility of these transformations is verified in Sect. 3.4. Here it turns out that the grading parameter has to be chosen sufficiently high. In view of the evolution step for the averages that has to be performed afterwards, it is helpful if the adaptive grid is locally uniform. Again, the grading plays an important role, see Sect. 3.5. Finally, we outline the algorithms by which the local transformations are performed and recall the core ingredients of the setting for curvilinear grids which are involved in the numerical experiments presented in Chapt. 7.

We would like to mention that locally refined spaces constructed by means of thresholded multiscale sequences are subject of current research arising in different fields, e.g. composite functions [DSX00] and multigrid [DMS02]. It turns out that the setting has always to be adapted to the problem at hand.

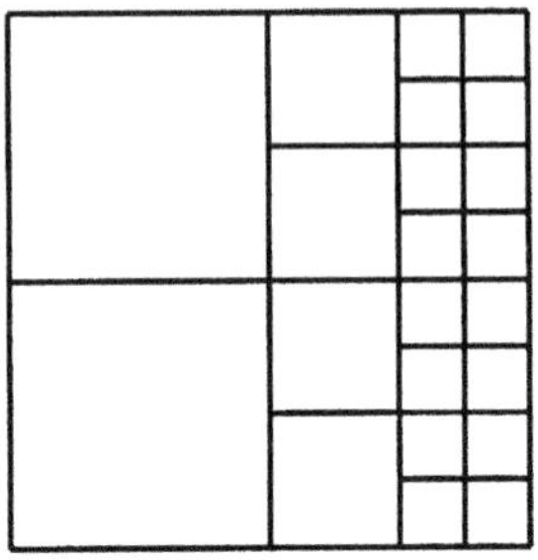

Fig. 3.1. Locally refined grid with hanging nodes

3.1 Adaptive Grid and Significant Details

We are now concerned with the construction of locally refined spaces. First of all, we introduce the notion of an adaptive grid by means of the nested grid hierarchy.

Definition 5. *(Adaptive Grid)*
Let $\mathcal{G}$ *be a nested grid hierarchy. Then an* adaptive grid *is determined by an index set* $\mathcal{G}_L \subset \bigcup_{j=0}^{L} \{j\} \times I_j$, *if the two properties*

- $\Omega = \bigcup_{(j,k)\in\mathcal{G}_L} V_{j,k}$,
- $|V_{j,k} \cap V_{j',k'}| = 0, \quad (j,k) \neq (j',k'),\ (j,k),(j',k') \in \mathcal{G}_L$

hold.

The grid corresponding to the set $\mathcal{G}_L$ can be interpreted as a locally refined grid with hanging nodes, i.e., cell vertices of neighboring cells do not necessarily coincide, see Fig. 3.1. In slight abuse of notation we will refer to the index set $\mathcal{G}_L$ as the adaptive grid.

An adaptive grid is determined by means of the multiscale sequence

$$\mathbf{d}^{(L)} = \left(\hat{\mathbf{u}}_0^T, \mathbf{d}_0^T, \ldots, \mathbf{d}_{L-1}^T\right)^T, \quad \mathbf{d}_j := (\mathbf{d}_{j,e})_{e\in E^*},\ \mathbf{d}_{j,e} := (d_{j,k,e})_{k\in I_j},$$

where the details $d_{j,k,e}$ correspond to the modified box wavelet $\tilde{\psi}_{j,k,e}$. This sequence is compressed by applying hard thresholding, i.e., all details $d_{j,k,e}$ smaller than a prescribed threshold value ε_j are discarded. The truncated sequence is determined by the index set

$$\mathcal{D}_{L,\varepsilon} := \{(j,k,e)\ ;\ |d_{j,k,e}| > \varepsilon_j,\ e\in E^*,\ k\in I_j,\ j\in\{0,\ldots,L-1\}\}, \quad (3.1)$$

where $\varepsilon = (\varepsilon_0,\ldots,\varepsilon_L)^T$ denotes a sequence of tolerances. Again, in slight abuse of notation we will refer to this index set as *set of significant details*. The corresponding details $(d_{j,k,e})_{(j,k,e)\in\mathcal{D}_{L,\varepsilon}}$ is called the *sequence of significant details.* In analogy, we introduce the *sequence of local averages* $(\hat{u}_{j,k})_{(j,k)\in\mathcal{G}_{L,\varepsilon}}$ corresponding to the adaptive grid.

The adaptive grid determined by the index set $\mathcal{G}_{L,\varepsilon}$ can now be constructed by means of the index set $\mathcal{D}_{L,\varepsilon}$ where we apply the following refinement criterion.

Definition 6. *(Refinement criterion)*
Let $\mathcal{G}$ be a nested grid hierarchy and $\mathcal{D}_{L,\varepsilon}$ a set of significant details. Then a cell $V_{j,k}$, $k \in I_j$, is refined if and only if there is $e \in E^*$ such that $(j,k,e) \in \mathcal{D}_{L,\varepsilon}$.

For later use we introduce the local index sets

$$I_{j,\varepsilon} := \{k \ ; \ (j,k) \in \mathcal{G}_{L,\varepsilon}\} \subset I_j, \ J_{j,\varepsilon} := \{(k,e) \ ; \ (j,k,e) \in \mathcal{D}_{L,\varepsilon}\} \subset I_j \times E^*,$$

which correspond to the indices of $\mathcal{G}_{L,\varepsilon}$ and $\mathcal{D}_{L,\varepsilon}$ on one level j.

The adaptive grid $\mathcal{G}_{L,\varepsilon}$ is now constructed by the following refinement algorithm.

Algorithm 2. *(Refinement Algorithm)*

1. *Initialize* $I_0^+ := I_0$;
2. for $j = 0$ to L-1 do
   1. *Initialize* $I_{j+1}^+ := \emptyset$, $I_j^- := \emptyset$;
   2. *Apply the refinement criterion, i.e.,* for $(k,e) \in J_{j,\varepsilon}$ do $I_j^- := I_j^- \cup \{k\}$; $I_{j+1}^+ := I_{j+1}^+ \cup \mathcal{M}_{j,k}$;
   3. *Discard the refined cells, i.e.,* $I_{j,\varepsilon} := I_j^+ \backslash I_j^-$.

Here the sets $I_j^+ \subset I_j$ and $I_j^- \subset I_j$ can be interpreted as collections of indices indicating cells on level j which might be refined on the next finer level when applying the refinement criterion and those which have been refined, respectively. Whenever a cell $V_{j,k}$ is refined, then the indices of the new cells $\mathcal{M}_{j,k} \subset I_{j+1}$ have to be added to I_{j+1}^+ and the index k of the refined cell has to be removed from I_j^+. The resulting grid is an adaptive grid in the sense of Definition 5. The grid refinement is illustrated schematically in Fig. 3.2 where the refined cells are shaded.

Finally, we introduce the locally refined spaces

$$\Phi_{L,\varepsilon} := \{\varphi_{j,k} \ ; \ (j,k) \in \mathcal{G}_{L,\varepsilon}\}, \quad \Psi_{L,\varepsilon} := \Phi_0 \cup \{\psi_{j,k,e} \ ; \ (j,k,e) \in \mathcal{D}_{L,\varepsilon}\}.$$

In analogy to (2.43), the projection u_L can now be represented in terms of both bases in the form

$$u_L = \sum_{(j,k)\in\mathcal{G}_{L,\varepsilon}} \hat{u}_{j,k}\,\varphi_{j,k} = \sum_{k\in I_0} \hat{u}_{0,k}\,\varphi_{0,k} + \sum_{(j,k,e)\in\mathcal{D}_{L,\varepsilon}} d_{j,k,e}\,\psi_{j,k,e}. \tag{3.2}$$

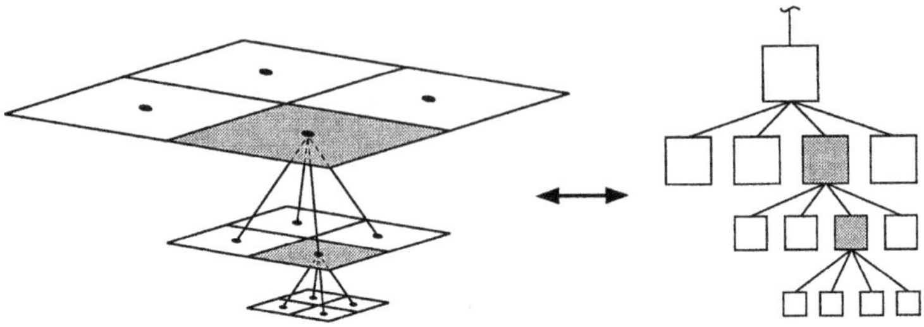

Fig. 3.2. Grid refinement

Note, that the cardinality of $\mathcal{D}_{L,\varepsilon}$ and $\mathcal{G}_{L,\varepsilon}$, respectively, might be much smaller than the number of cells of the finest grid, i.e.,

$$\#\mathcal{D}_{L,\varepsilon} \ll N_L$$

which, of course, depends on the regularity of u. In order to switch between the single–scale and the wavelet representation of u_L according to (3.2) we will apply a *local* multiscale transformation analogously to (2.40). This transformation is to be realized with an optimal complexity, i.e., the number of operations should stay proportional to $\#\mathcal{D}_{L,\varepsilon}$.

3.2 Grading

In order to realize the local multiscale transformation in *one* sweep through the refinement levels and in view of its feasibility we have to inflate the set of significant details $\mathcal{D}_{L,\varepsilon}$ by a grading procedure. For this purpose, we first agglomerate all wavelet indices corresponding to a cell $V_{j,k}$, $k \in I_j$, i.e.,

$$T_{j,k} := \{(k,e) \; ; \; e \in E^*\}$$

and introduce the stencil of neighbors $\mathcal{N}_{j,k} \subset I_j$ corresponding to $V_{j,k}$ by the recursive definition

$$\begin{aligned} \mathcal{N}^0_{j,k} &:= \{k\}, \\ \mathcal{N}^i_{j,k} &:= \{r \in I_j \; : \; \exists\, s \in \mathcal{N}^{i-1}_{j,k} \text{ s.t. } V_{j,r} \cap V_{j,s} \neq \emptyset\}, \; i = 1, \ldots, q. \end{aligned} \tag{3.3}$$

Note, that all cells are assumed to be closed. Then $\mathcal{N}^q_{j,k}$ is referred to as the *neighborhood of degree* $q \in \mathbb{N}_0$. This set is the union of neighboring cells corresponding to $V_{j,k}$ which are attached to each other by at least one point. Hence the support $\bigcup_{l \in \mathcal{N}^q_{j,k}} V_{j,l}$ is a simply connected domain with no holes provided this holds also for the computational domain Ω. Obviously, the following relations hold

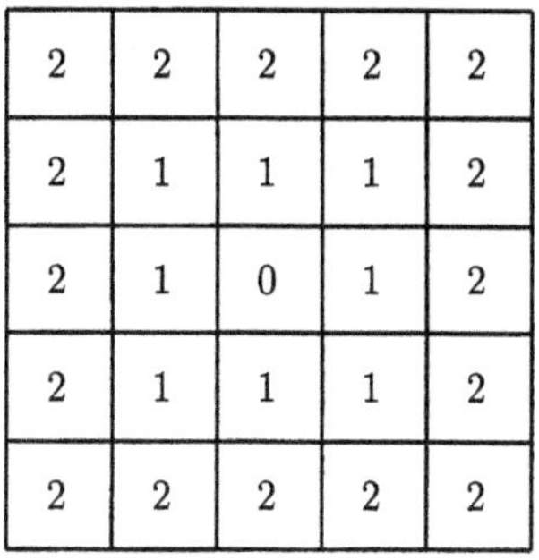

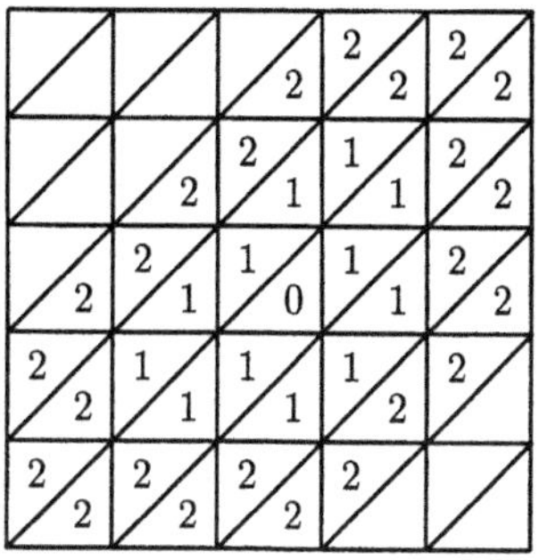

Fig. 3.3. Illustration of the sets $\mathcal{N}^q_{j,k}$ for $q = 0, 1, 2$ and $d = 2$

$$l \in \mathcal{N}^q_{j,k} \Leftrightarrow k \in \mathcal{N}^q_{j,l} \quad \text{and} \quad \mathcal{N}^q_{j,k} \subset \mathcal{N}^{q+1}_{j,k}. \tag{3.4}$$

In Fig. 3.3 the sets $\mathcal{N}^q_{j,k}$ are illustrated for two–dimensional grids composed of quadrilaterals and triangles, respectively. Here the cell $V_{j,k}$ is indicated by 0. The neighborhood of degree q is determined by all cells labeled with the numbers $q, q-1, \ldots, 0$.

By the sets $\mathcal{M}_{j,k} \subset I_{j+1}$ we know the indices of the new cells generated by refining the cell $V_{j,k}$. Sometimes it is helpful to know the inverse relation, i.e., we are interested in the index k of the coarse cell corresponding to $V_{j+1,r}, r \in \mathcal{M}_{j,k}$. This information is provided by the operator $\pi_{j+1} : I_{j+1} \to I_j,\ j = 0, \ldots, L-1$, defined by

$$\pi_{j+1}(r) = k \quad \text{with} \quad r \in \mathcal{M}_{j,k}. \tag{3.5}$$

Note, that this function is well–defined, since the nested grid hierarchy according to Definition 3 is assumed to be gap– and redundancy–free. Hence, for any $r \in I_{j+1}$ there is exactly one index $k \in I_j$ such that $r \in \mathcal{M}_{j,k}$. For instance, in the one–dimensional case the refinement set corresponding to the cell $V_{j,k}$ is given by $\mathcal{M}_{j,k} = \{2k, 2k+1\}$ according to the uniform dyadic grid refinement, cf. Sects. 2.2 and 2.5.2. Hence, $\pi_{j+1}(r) = \lfloor r/2 \rfloor = k$ for $r \in \mathcal{M}_{j,k}$.

By means of the above settings we can now introduce the notion of a graded tree.

Definition 7. *(Graded Tree)*
The set of significant details $\mathcal{D}_{L,\varepsilon}$ is called a graded tree of degree q, *if for any $j \in \{1, \ldots, L\}$ the relation*

$$(k, e) \in J_{j,\varepsilon} \Rightarrow T_{j-1,r} \subset J_{j-1,\varepsilon}, \quad \forall\, r \in \mathcal{N}^q_{j-1,\pi_j(k)} \tag{3.6}$$

holds.

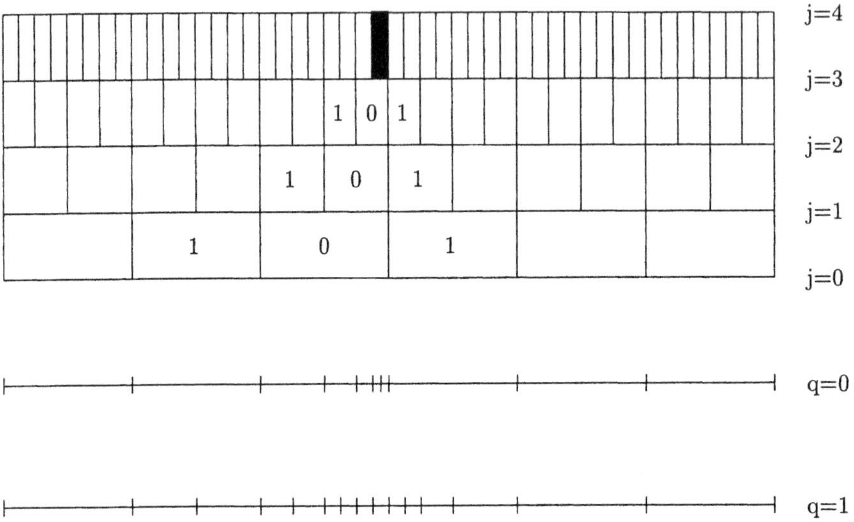

Fig. 3.4. Graded trees of degree $q = 0, 1$ (top) and corresponding adaptive grids $q = 0$ (middle), $q = 1$ (bottom)

The grading of the truncated details results in a tree structure of the details. Generally speaking, a graded tree means that for any significant detail on level j there are significant details in the neighborhood but on the next lower level $j - 1$. Thus, gradedness simplifies the local multiscale transformation, since there are no isolated details on higher levels. In addition, locally switching between finer and coarser levels becomes easier, because only two levels are involved if the grading parameter q is chosen sufficiently large. In particular, the level difference between two neighboring cells is at most 1. Hence, we have to look for all neighbors only on the same level, the next coarser level or the next finer level, i.e., we avoid recursive searching in the tree. We emphasize that for $q = 0$ the details still correspond to a tree but not necessarily a graded tree. In particular, a graded tree of degree q is also a graded tree of degree $q' < q$.

An example is presented in Fig. 3.4 for the one–dimensional case. Here the coarsest uniform grid is composed of six intervals which are successively refined by dyadic grid refinement. We assume that there is one significant detail on refinement level 3 determined by $(j, k) = (3, 23)$. For the univariate setting, cf. Sects. 2.2 and 2.5.2, Definition 7 implies for $k \in J_{j,\varepsilon}$

$$\mathcal{N}^q_{j-1,\lfloor k/2 \rfloor} \equiv \{\max(0, \lfloor k/2 \rfloor - q), \ldots, \min(\lfloor k/2 \rfloor + q, N_j - 1)\} \subset J_{j-1,\varepsilon}$$

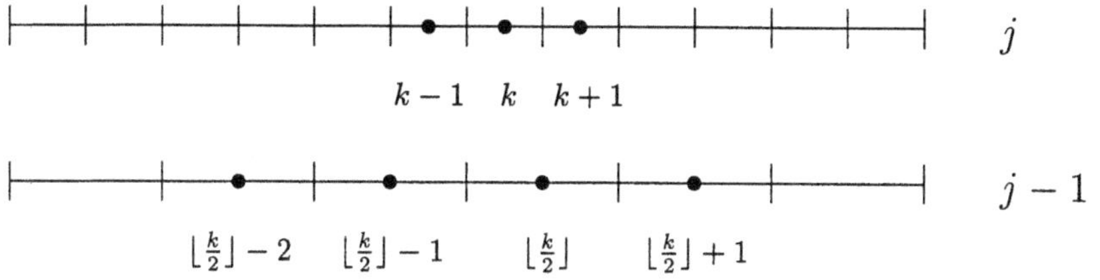

Fig. 3.5. Nestedness of neighborhoods of degree $q = 1$

where we omit the index e for the wavelet type ($e = 1$). Then the graded tree of degree 0 is determined by all intervals with label "0" whereas the graded tree of degree 1 is determined by all intervals with label "0" and "1". For both cases, $q = 0$ and $q = 1$, the corresponding adaptive grid according to Algorithm 2 are also presented in Fig. 3.4. Note that for $q = 1$ two neighboring intervals differ at most by one refinement level. Obviously, this is not true for $q = 0$.

In the following we investigate the effect of grading the significant details on the adaptive grid. For this purpose, we first observe that the nestedness of the underlying grid hierarchy implies the nestedness of the neighborhoods. This conclusion is illustrated in Fig. 3.5 for the one–dimensional case.

Lemma 1. *(Nestedness of neighborhoods)*
If the grid hierarchy is nested, then the supports of the neighborhoods are also nested, i.e.,

$$\bigcup_{l \in \mathcal{N}^q_{j,k}} V_{j,l} \subset \bigcup_{l \in \mathcal{N}^q_{j-1,\pi_j(k)}} V_{j-1,l}. \tag{3.7}$$

Proof: The assertion is proved by induction over q. For $q = 0$ we know by initialization of the neighborhoods that $\mathcal{N}^0_{j,k} = \{k\}$ and $\mathcal{N}^0_{j-1,\pi_j(k)} = \{\pi_j(k)\}$. Hence, the nestedness of the grid hierarchy implies the assertion in this case. We now assume that the assertion holds for some q and decompose

$$\bigcup_{l \in \mathcal{N}^{q+1}_{j,k}} V_{j,l} = \bigcup_{l \in \mathcal{N}^{q+1}_{j,k} \setminus \mathcal{N}^q_{j,k}} V_{j,l} \cup \bigcup_{l \in \mathcal{N}^q_{j,k}} V_{j,l}.$$

The induction assumption together with the recursive definition of the neighborhoods and (3.4) imply

$$\bigcup_{l \in \mathcal{N}^q_{j,k}} V_{j,l} \subset \bigcup_{l \in \mathcal{N}^q_{j-1,\pi_j(k)}} V_{j-1,l} \subset \bigcup_{l \in \mathcal{N}^{q+1}_{j-1,\pi_j(k)}} V_{j-1,l}.$$

By definition of $\mathcal{N}_{j,k}^{q+1}$ the additional layer of cells is attached to the neighborhood $\mathcal{N}_{j,k}^{q}$ which is also true for $\mathcal{N}_{j-1,\pi_j(k)}^{q+1}$ and $\mathcal{N}_{j-1,\pi_j(k)}^{q}$. Again we conclude from the nestedness of the grid hierarchy and the induction assumption

$$\{\pi_j(l) \ : \ l \in \mathcal{N}_{j,k}^{q+1} \backslash \mathcal{N}_{j,k}^{q}\} \subset \mathcal{N}_{j-1,\pi_j(k)}^{q+1}.$$

Thus, we obtain (3.7) with q replaced by $q+1$ which completes the proof. □

From this lemma we deduce the following properties.

Corollary 3.
If the grid hierarchy is nested, then the following relations hold:

1.) $\mathcal{N}_{j,k}^{q} \subset \bigcup_{l \in \mathcal{N}_{j-1,\pi_j(k)}^{q}} \mathcal{M}_{j-1,l}$;

2.) $\bigcup_{l \in \mathcal{N}_{j,k}^{q}} V_{j-1,\pi_j(l)} \subset \bigcup_{l \in \mathcal{N}_{j-1,\pi_j(k)}^{q}} V_{j-1,l}$ *or equivalently*

$$\pi_j\left(\mathcal{N}_{j,k}^{q}\right) := \bigcup_{l \in \mathcal{N}_{j,k}^{q}} \pi_j(l) \subset \mathcal{N}_{j-1,\pi_j(k)}^{q}.$$

Proof. **1.)** The assertion follows by Lemma 1 and the nestedness of the cells since

$$\bigcup_{l \in \mathcal{N}_{j,k}^{q}} V_{j,l} \subset \bigcup_{l \in \mathcal{N}_{j-1,\pi_j(k)}^{q}} V_{j-1,l} = \bigcup_{l \in \mathcal{N}_{j-1,\pi_j(k)}^{q}} \bigcup_{r \in \mathcal{M}_{j-1,l}} V_{j,r}$$

holds.

2.) On account of property 1, the definition of π_j and the nestedness of the grid hierarchy we conclude

$$\bigcup_{l \in \mathcal{N}_{jk}^{q}} V_{j-1,\pi_j(l)} \subset \bigcup_{l \in \mathcal{N}_{j-1,\pi_j(k)}^{q}} \bigcup_{r \in \mathcal{M}_{j-1,l}} V_{j-1,\pi_j(r)} =$$

$$\bigcup_{l \in \mathcal{N}_{j-1,\pi_j(k)}^{q}} \bigcup_{r \in \mathcal{M}_{j-1,l}} V_{j-1,l} = \bigcup_{l \in \mathcal{N}_{j-1,\pi_j(k)}^{q}} V_{j-1,l}.$$

This yields the assertion. □

The structure of the adaptive grid can now be characterized as follows.

Lemma 2. *(Λ–structure of adaptive grid)*
Assume that the tree corresponding to the significant details is graded of degree $q \geq 0$. Then the adaptive grid constructed by Algorithm 2 is Λ–structured, i.e., for any $k \in I_{j+1,\varepsilon}$ the following two relations hold:

1.) $\pi_i \circ \cdots \circ \pi_{j+1}(k) \notin I_{i-1,\varepsilon}, \quad i = 1, \ldots, j+1$;

2.) $T_{i-2,\pi_{i-1}\circ\cdots\circ\pi_{j+1}(k)} \subset J_{i-2,\varepsilon}, \quad i = 2, \ldots, j+1.$

Proof. For simplicity we define $k_i := \pi_i \circ \cdots \circ \pi_{j+1}(k)$. The assertion is proven by induction over i. For $i = j+1$ we conclude from the refinement criterion, see Definition 6, that there is some index $(k_{j+1}, e) \in J_{j,\varepsilon}$. Therefore $V_{j,k_{j+1}}$ is refined according to Algorithm 2. Furthermore, the assumption of a graded tree implies that $T_{j-1,r} \subset J_{j-1,\varepsilon}$ for all $r \in \mathcal{N}^q_{j-1,\pi_j(k_{j+1})}$. Since $\pi_j(k_{j+1}) = k_j \in \mathcal{N}^0_{j-1,k_j} \subset \mathcal{N}^q_{j-1,k_j}$ we obtain, in particular, for $r = k_j$ that $T_{j-1,k_j} \subset J_{j-1,\varepsilon}$.

We now assume that the assertion holds for some $i \geq 2$. Then the gradedness yields $T_{i-3,r} \subset J_{j-3,\varepsilon}$ for all $r \in \mathcal{N}^q_{i-3,\pi_{i-2}(k_{i-1})}$. Since $\pi_{i-2}(k_{i-1}) = k_{i-2} \in \mathcal{N}^0_{i-3,k_{i-2}} \subset \mathcal{N}^q_{i-3,k_{i-2}}$ we conclude for $r = k_{i-2}$ that $T_{i-3,k_{i-2}} \subset J_{i-3,\varepsilon}$. Furthermore, the assumption $k_{i-1} \in I_{i-2,\varepsilon}$ would lead to a contradiction to the induction assumption, since the refinement criterion implies $T_{i-2,k_{i-1}} \not\subset J_{i-2,\varepsilon}$. This completes the proof. □

This lemma shows how the grading of the set of significant details influences the adaptive grid. In particular, the gradedness of the tree corresponding to the significant details results in a Λ–structure of the adaptive grid. Thus the adaptive grid can be interpreted as the leaves of the tree. An example is shown in Fig. 3.6 illustrating these statements in the one–dimensional case. Here we consider a piecewise constant function u on the unit interval $[0,1]$. The multiscale analysis is performed on a uniform dyadic grid hierarchy, cf. Sect. 2.2, using the modified box wavelets according to Sect. 2.5.2 with $s = 2$ and grading of degree $q = 2$. The index sets $\mathcal{G}_{L,\varepsilon}$ and $\mathcal{D}_{L,\varepsilon}$ are represented by plotting the position k at the cell center of an interval $V_{j,k}$ versus the level j. The function u is represented by the averages $\hat{u}_{j,k}$ plotted at the cell center corresponding to the adaptive grid.

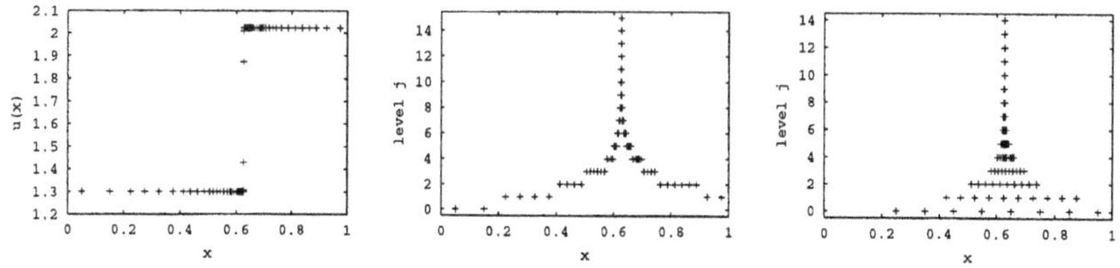

Fig. 3.6. Averages of function u (left), distribution of $\mathcal{G}_{L,\varepsilon}$ (middle), distribution of $\mathcal{D}_{L,\varepsilon}$ (right)

We are now able to verify that the grading of the details implies the grading of the adaptive grid, i.e., two neighboring cells of the adaptive grid differ at most by one level.

Proposition 1. *(Graded tree ⇒ graded grid)*
Assume that the tree of significant details is graded of degree $q \geq 1$. Then the corresponding adaptive grid determined by Algorithm 2 is also graded of

degree q, i.e., for any $k \in I_{j,\varepsilon}$ and $l \in \mathcal{N}^q_{j,k}$ one of the following three cases holds:

1.) $l \in I_{j,\varepsilon}$ or

2.) $l \notin I_{j,\varepsilon}$ and $\pi_j(l) \in I_{j-1,\varepsilon}$ or

3.) $l \notin I_{j,\varepsilon}$ and $\mathcal{M}_{j,l} \subset I_{j+1,\varepsilon}$.

Proof. Let $k \in I_{j,\varepsilon}$. Then we conclude from the refinement criterion, see Definition 6, that there exists $(\pi_j(k), e) \in J_{j-1,\varepsilon}$ and, hence, the property (3.6) of a graded tree implies $T_{j-2,r} \subset J_{j-2,\varepsilon}$ for all $r \in \mathcal{N}^q_{j-2,\pi_{j-1}(\pi_j(k))}$. Furthermore, Algorithm 2 guarantees that $\mathcal{M}_{j-2,r} \subset I^+_{j-1}$ for these indices r. We obtain now by Corollary 3 the inclusion

$$\pi_j\left(\mathcal{N}^q_{j,k}\right) \subset \mathcal{N}^q_{j-1,\pi_j(k)} \subset \bigcup_{r\in\mathcal{N}^q_{j-2,\pi_{j-1}(\pi_j(k))}} \mathcal{M}_{j-2,r}.$$

From this relation we conclude that $\pi_j(l) \in I^+_{j-1}$ for all $l \in \mathcal{N}^q_{j,k}$. According to Algorithm 2 there are two possibilities, either $\pi_j(l) \in I_{j-1,\varepsilon}$ or $V_{j-1,\pi_j(l)}$ has been refined, i.e., $l \in I^+_j$. In the latter case we have to check whether the cell $V_{j,l}$ has been refined again but at most once. For this purpose, we assume that the cell $V_{j,l}$ for an arbitrary $l \in \mathcal{N}^q_{j,k}$ has been refined twice, i.e., there is $s \in I^+_{j+2}$ such that $\pi_{j+1}(\pi_{j+2}(s)) = l$. According to the refinement criterion, this can only be the case if $(\pi_{j+2}(s), e) \in J_{j+1,\varepsilon}$ for an $e \in E^*$. Then the gradedness of the details implies $T_{j,r} \subset J_{j,\varepsilon}$ for all $r \in \mathcal{N}^q_{j,\pi_{j+1}(\pi_{j+2}(s))} = \mathcal{N}^q_{j,l}$. Since $l \in \mathcal{N}^q_{j,k}$ we conclude by Definition 3.3 of the neighborhoods that $k \in \mathcal{N}^q_{j,k}$ and, hence, $(k, e) \in J_{j,\varepsilon}$. In view of the refinement criterion the cell $V_{j,k}$ has to be refined, i.e., $k \notin I_{j,\varepsilon}$. This is a contradiction to the assumption. □

From the conditions 1.), 2.) and 3.) of Proposition 1 we conclude that an adaptive grid is graded of degree q if for any cell $V_{j,k}$ of the adaptive grid all cells corresponding to the neighborhood $\mathcal{N}^q_{j,k}$ of degree q is (1) itself a cell of the adaptive grid or either the (2) father or (3) the children are cells of the adaptive grid. This is schematically illustrated in Fig. 3.7 for $q = 2$. Here the cell $V_{j,k}$ and the cells of the neighborhood $\mathcal{N}^q_{j,k}$ are indicated by ∘ and •, respectively. At the bottom of Fig. 3.7 one possible adaptive grid is shown that is graded of degree 2. In Sect. 4.1.1 the gradedness of the adaptive grid will be helpful for performing the flux computation efficiently on a locally uniform grid.

Conversely, the grading of the grid implies the grading of the tree corresponding to the significant details.

Proposition 2. *(Graded grid ⇒ graded tree)*
Assume that the adaptive grid is graded of degree p. If the condition

$$\mathcal{N}^q_{j-1,\pi_j(k)} \subset \pi_j(\pi_{j+1}(\mathcal{N}^p_{j+1,k'})) \tag{3.8}$$

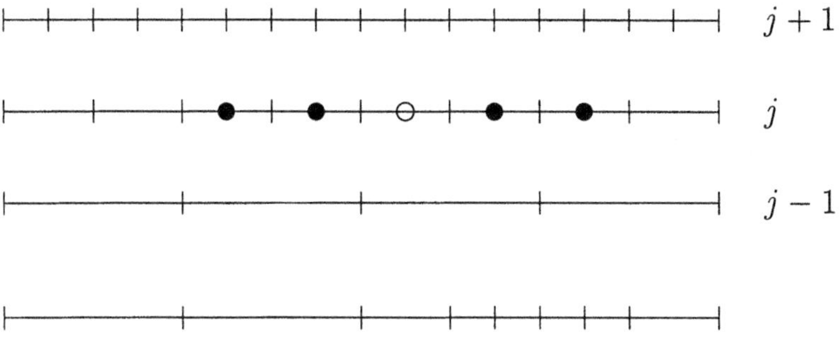

Fig. 3.7. Graded adaptive grid of degree $q = 2$

holds for any $k \in I_j$ with $T_{j,k} \subset J_{j,\varepsilon}$ and $k' \in \mathcal{M}_{j,k}$, then the tree of the details is graded of degree q.

Proof. Let $k \in I_j$ with $T_{j,k} \subset J_{j,\varepsilon}$. Then the grading of the details implies

$$T_{j-1,r} \subset J_{j-1,\varepsilon} \qquad \forall r \in \mathcal{N}^q_{j-1,\pi_j(k)}. \tag{3.9}$$

On the other hand we conclude from the Proposition 1 that for all $k' \in \mathcal{M}_{j,k}$ and $l \in \mathcal{N}^p_{j+1,k'}$ one of the three conditions holds (i) $l \in I_{j+1,\varepsilon}$ or (ii) $l \notin I_{j+1,\varepsilon}$ and $\pi_{j+1}(l) \in I_{j,\varepsilon}$ or (iii) $l \notin I_{j+1,\varepsilon}$ and $\mathcal{M}_{j+1,l} \subset I_{j+2,\varepsilon}$. The refinement criterion implies that $\pi_{j+1}(\mathcal{N}^p_{j+1,k'}) \subset I^+_j$. According to the refinement algorithm this holds only if

$$\pi_j(\pi_{j+1}(\mathcal{N}^p_{j+1,k'})) \subset J_{j-1,\varepsilon}. \tag{3.10}$$

Comparing (3.9) and (3.10) the assertion follows immediately. □

Obviously, condition (3.8) is always satisfied for $q = 0$, since $\pi_{j+1}(k') = k$ and $k' \in \mathcal{N}^p_{j+1,k'}$ and, hence,

$$\mathcal{N}^0_{j-1,\pi_j(k)} = \{\pi_j(k)\} = \{\pi_j(\pi_{j+1}(k'))\} \subset \pi_j(\pi_{j+1}(\mathcal{N}^p_{j+1,k'})).$$

Furthermore, condition (3.8) gives an upper bound for the grading parameter q depending on p.

In Fig. 3.8 the condition (3.8) is illustrated for the one–dimensional case according to Sect. 2.5.2. Hence the uniform dyadic grid hierarchy implies $k' \in \mathcal{M}_{j,k} = \{2k, 2k+1\}$. According to (3.5) and the dyadic grid refinement, cf. Sects. 2.2 and 2.5.2, we determine the index sets of the neighborhoods $\mathcal{N}^p_{j+1,k'} = \{k'-p, \ldots, k'+p\} \subset I_{j+1}$ and $\mathcal{N}^q_{j-1,\pi_j(k)} = \{\lfloor k/2 \rfloor - q, \ldots, \lfloor k/2 \rfloor + q\} \subset I_{j-1}$ as well as the index set of the projection $\pi_{j+1}(\mathcal{N}^p_{j+1,k'}) = \{\lfloor (k'-p)/2 \rfloor, \ldots, \lfloor (k'+p)/2 \rfloor\} \subset I_j$. These sets are indicated by • and ∘, respectively. From Fig. 3.8 it can be motivated that the condition (3.8) holds if $0 \le q \le \lfloor p/4 \rfloor$, cf. Corollary 6 in Sect. 3.8.

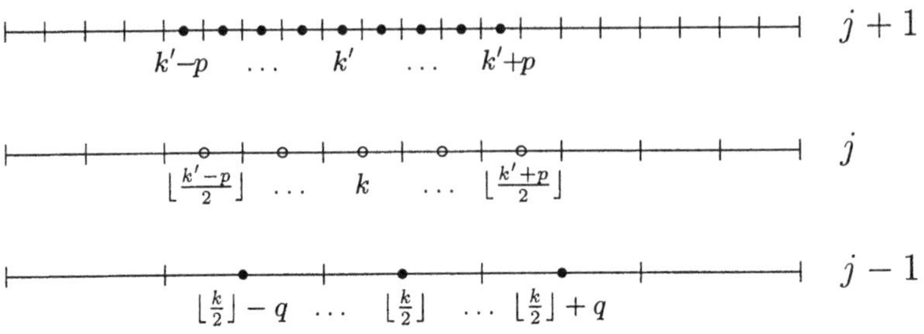

Fig. 3.8. Illustration of condition (3.8)

3.3 Local Multiscale Transformation

So far we have introduced locally refined spaces whose dimension corresponds to $\#\mathcal{D}_{L,\varepsilon}$ and $\#\mathcal{G}_{L,\varepsilon}$, respectively. Now we have to explain how to perform the change of bases between the single–scale and the wavelet representation in (3.2) with a number of operations proportional to $\#\mathcal{D}_{L,\varepsilon}$. This is only possible if the involved matrices are uniformly banded and, hence, the mask matrices are sparse, see Sect. 2.6. In the sequel we proceed in analogy to the transformation (2.40) of the full spaces. For this purpose we first summarize the multiscale transformation for the averages and details defined by (2.3) and (2.26). From (2.3), (2.26) and (2.29), (2.30) we first derive the two–scale relations

$$\hat{\mathbf{u}}_j = \check{\mathsf{M}}_{j,0}^T \, \hat{\mathbf{u}}_{j+1}, \qquad \check{\mathbf{d}}_{j,e} = \check{\mathsf{M}}_{j,e}^T \, \hat{\mathbf{u}}_{j+1}, \quad e \in E^* \tag{3.11}$$

$$\hat{\mathbf{u}}_{j+1} = \check{\mathsf{G}}_{j,0}^T \, \hat{\mathbf{u}}_j + \sum_{e \in E^*} \check{\mathsf{G}}_{j,e}^T \, \check{\mathbf{d}}_{j,e}, \tag{3.12}$$

where the matrices of the mask coefficients are determined by (2.31).

For the modified box wavelets similar relations hold where the mask matrices $\check{\mathsf{M}}_{j,e}$, $\check{\mathsf{G}}_{j,e}$ are replaced by $\widetilde{\mathsf{M}}_{j,e}$, $\widetilde{\mathsf{G}}_{j,e}$ determined by (2.37) and (2.38). According to (2.33) these matrices are related by a modification matrix $\mathsf{L}_{j,e}$ determined by

$$(\mathsf{L}_{j,e})_{k,r} = \begin{cases} l_{k,l}^{j,e} & , r \in \mathcal{M}_{j,l}, \ l \in \mathcal{L}_{j,k}^e \cup \{k\}, \\ 0 & , \text{elsewhere}, \end{cases}$$

for $r \in I_{j+1}$, $k \in I_j$, corresponding to the coarse grid modification of the box wavelets. To obtain an efficient multiscale transformation the computation of the modified matrices $\widetilde{\mathsf{M}}_{j,e}$ and $\widetilde{\mathsf{G}}_{j,e}$ according to (2.33) should be avoided. Instead it is more efficient to apply the right hand sides directly to the vectors of the averages and the details, respectively, because this requires less operations. Then the two–scale relations for the multiscale transformation, i.e., $\hat{\mathbf{u}}_j$ and $\mathbf{d}_{j,e}$ are computed by $\hat{\mathbf{u}}_{j+1}$, read

$$\begin{aligned} \hat{\mathbf{u}}_j &= \check{\mathsf{M}}_{j,0}^T\, \hat{\mathbf{u}}_{j+1}, \\ \mathbf{d}_{j,e} &= \check{\mathbf{d}}_{j,e} + \mathbf{w}_{j,e},\ \ \mathbf{w}_{j,e} := \mathsf{L}_{j,e}\, \hat{\mathbf{u}}_j,\ \ \check{\mathbf{d}}_{j,e} := \check{\mathsf{M}}_{j,e}^T\, \hat{\mathbf{u}}_{j+1},\ \ e \in E^*. \end{aligned} \tag{3.13}$$

Here the vectors $\mathbf{w}_{j,e}$ and $\check{\mathbf{d}}_{j,e}$ denote the coarse grid correction of the box wavelets and the details of the box wavelets. Analogously, the inverse multiscale transformation, i.e., $\hat{\mathbf{u}}_{j+1}$ is computed by $\hat{\mathbf{u}}_j$ and $\mathbf{d}_{j,e}$, reads

$$\begin{aligned} \hat{\mathbf{u}}_{j+1} &= \check{\mathsf{G}}_{j,0}^T\, \hat{\mathbf{u}}_j + \sum_{e \in E^*} \check{\mathsf{G}}_{j,e}^T\, \check{\mathbf{d}}_{j,e}, \\ \check{\mathbf{d}}_{j,e} &:= \mathbf{d}_{j,e} - \mathbf{w}_{j,e},\ \ \mathbf{w}_{j,e} := \mathsf{L}_{j,e}\, \hat{\mathbf{u}}_j,\ \ e \in E^*. \end{aligned} \tag{3.14}$$

In order to realize the change of basis between the single–scale representation and the wavelet representation in (3.2) with an effort that is proportional to the number of significant details $\#\,\mathcal{D}_{L,\varepsilon}$ and local averages $\#\,\mathcal{G}_{L,\varepsilon}$, respectively, it is prohibited to access the full coefficient vectors and mask matrices. Instead we have to carry out the transformations (3.13) and (3.14) componentwise only for those coefficients that correspond to $\mathcal{G}_{L,\varepsilon}$ and $\mathcal{D}_{L,\varepsilon}$. In particular, the summation is restricted to those indices which correspond to non–vanishing entries of the mask matrices. In order to describe this local transformation the notion of the *support* of matrix columns and rows is helpful, i.e.,

$$\mathcal{A}_k := \mathrm{supp}(\mathsf{A}, k) := \{r\ ;\ a_{r,k} \neq 0\} = \text{ support of } k\text{th column of } \mathsf{A},$$

$$\mathcal{A}_k^* := \mathrm{supp}(\mathsf{A}^T, k) := \{r\ ;\ a_{k,r} \neq 0\} = \text{ support of } k\text{th row of } \mathsf{A}.$$

These sets are only helpful when the matrix A is sparse, i.e., only a few matrix entries are different from zero. In this case it is useful to know the non–trivial terms of the matrix–vector product $\mathbf{y}^T = \mathbf{x}^T \mathsf{A}$ that have to be computed. The support $\mathcal{A}_k$ of a column collects all non–vanishing matrix elements that might yield a non–trivial contribution to the kth component of the matrix–vector product, i.e., y_k. Therefore it can be interpreted as the *domain of dependence* for y_k, i.e., the components x_r which contribute to y_k. The support $\mathcal{A}_k^*$ of a row collects all non–vanishing matrix entries of the kth row that might yield a non–trivial contribution to the vector $\mathbf{y}$ of the matrix–vector product. Therefore it can be interpreted as the *range of influence*, i.e., the components y_r which are influenced by the component x_k. Obviously, the supports satisfy the relations

$$r \in \mathcal{A}_k \Rightarrow k \in \mathcal{A}_r^* \qquad \text{and} \qquad k \in \mathcal{A}_r^* \Rightarrow r \in \mathcal{A}_k .$$

For the description of the local multiscale transformation and its analytical investigation the supports of the box wavelets

- $\breve{\mathcal{G}}^e_{j,k} := \mathrm{supp}(\breve{\mathsf{G}}_{j,e}, k)$, $\breve{\mathcal{M}}^e_{j,k} := \mathrm{supp}(\breve{\mathsf{M}}_{j,e}, k)$, $e \in E$
- $\breve{\mathcal{G}}^{*,e}_{j,k} := \mathrm{supp}(\breve{\mathsf{G}}^T_{j,e}, k)$, $\breve{\mathcal{M}}^{*,e}_{j,k} := \mathrm{supp}(\breve{\mathsf{M}}^T_{j,e}, k)$, $e \in E$

and the supports of the modified box wavelets

- $\mathcal{L}^e_{j,k} := \mathrm{supp}(\mathsf{L}_{j,e}, k)$, $\mathcal{L}^{*,e}_{j,k} := \mathrm{supp}(\mathsf{L}^T_{j,e}, k)$, $e \in E^*$
- $\mathcal{G}^0_{j,k} := \breve{\mathcal{G}}^0_{j,k} \cup \mathcal{L}^e_{j,k}$, $\mathcal{G}^e_{j,k} := \breve{\mathcal{G}}^e_{j,k}$, $e \in E^*$
- $\mathcal{M}^0_{j,k} := \breve{\mathcal{M}}^0_{j,k} \equiv \mathcal{M}_{j,k}$, $\mathcal{M}^e_{j,k} := \breve{\mathcal{M}}^0_{j,k} \equiv \mathcal{M}_{j,k} \cup \mathcal{L}^e_{j,k}$, $e \in E^*$
- $\mathcal{G}^{*,0}_{j,k} := \breve{\mathcal{G}}^{*,0}_{j,k} \cup \mathcal{L}^{*,e}_{j,k}$, $\mathcal{G}^{*,e}_{j,k} := \breve{\mathcal{G}}^{*,e}_{j,k}$, $e \in E^*$
- $\mathcal{M}^{*,0}_{j,k} := \breve{\mathcal{M}}^{*,0}_{j,k}$, $\mathcal{M}^{*,e}_{j,k} := \breve{\mathcal{M}}^{*,0}_{j,k} \cup \mathcal{L}^{*,e}_{j,k}$, $e \in E^*$

will be needed. Note, that the supports $\mathcal{G}^e_{j,k}$, $\mathcal{M}^e_{j,k}$, $\mathcal{G}^{*,e}_{j,k}$ and $\mathcal{M}^{*,e}_{j,k}$ are not indicated by a tilde. In particular, the supports of the box wavelets are determined by

$$\breve{\mathcal{M}}^e_{j,k} = \breve{\mathcal{G}}^{*,e}_{j,k} = \breve{\mathcal{M}}^0_{j,k} \equiv \mathcal{M}_{j,k},\ \breve{\mathcal{M}}^{*,e}_{j,k} = \breve{\mathcal{G}}^e_{j,k} = \breve{\mathcal{M}}^{*,0}_{j,k},\ e \in E,\ k \in I_j \quad (3.15)$$

where the support $\breve{\mathcal{M}}^0_{j,k}$ coincides with the refinement set $\mathcal{M}_{j,k}$ corresponding to the cell $V_{j,k}$, cf. Sect. 2.1. With the aid of this notation we now introduce the local multiscale transformation considering (3.13) and (3.14) componentwise. For this purpose we have to specify which data are accessed when *locally* performing the transformation, i.e., we only consider the transformation for a single coefficient. Choose $k \in I_j$ such that $\breve{\mathcal{M}}^0_{j,k} \subset I_{j+1,\varepsilon}$. Then the *local two–scale transformation* for the k-th component of (3.13) reads

$$\hat{u}_{j,k} = \sum_{r \in \breve{\mathcal{M}}^0_{j,k}} \breve{m}^{j,0}_{r,k} \hat{u}_{j+1,r}, \quad (3.16)$$

$$w_{j,r,e} = \sum_{s \in \mathcal{L}^e_{j,r}} l^{j,e}_{s,r} \hat{u}_{j,s}, \quad r \in S^e_{j,k}, \quad (3.17)$$

$$\hat{u}_{j+1,s} = \sum_{r \in \breve{\mathcal{G}}^0_{j,s}} \breve{g}^{j,0}_{r,s} \hat{u}_{j,r} - \sum_{e \in E^*} \sum_{r \in \breve{\mathcal{G}}^e_{j,s}} \breve{g}^{j,e}_{r,s} w_{j,r,e}, \quad s \in \bigcup_{e \in E^*} P^e_{j+1,k}, \quad (3.18)$$

$$\breve{d}_{j,r,e} = \sum_{s \in \breve{\mathcal{M}}^e_{j,r}} \breve{m}^{j,e}_{s,r} \hat{u}_{j+1,s}, \quad r \in U^e_{j,k}, \quad (3.19)$$

$$d_{j,r,e} = \breve{d}_{j,r,e} + w_{j,r,e} = \sum_{s \in \mathcal{M}^e_{j,r}} \tilde{m}^{j,e}_{s,r} \hat{u}_{j+1,s}, \quad r \in U^e_{j,k}. \quad (3.20)$$

Here the index sets $U^e_{j,k} \subset I_j$, $P^e_{j+1,k} \subset I_{j+1}$ and $S^e_{j,k} \subset I_j$ are not yet specified. The main task of this chapter will be to derive these index sets such that each step of the local transformation can be carried out, i.e., all data on the right hand side have already been computed. In the proof of Lemma 3 they will be identified as

$$U^e_{j,k} = \mathcal{L}^{*,e}_{j,k}, \quad P^e_{j+1,k} = \bigcup_{r \in \mathcal{L}^{*,e}_{j,k} \setminus \{k\}} \check{\mathcal{M}}^0_{j,r}, \quad S^e_{j,k} = \mathcal{L}^{*,e}_{j,k}$$

according to the definitions (3.27), (3.28) and (3.29). Note, that we compute in (3.18) a prediction value for $\hat{u}_{j+1,s}$ which is accessed in (3.20). This prediction value has to be computed only if $s \notin I_{j+1,\varepsilon}$, or, equivalently, $d_{j,r,e} = 0$ for all coefficients $r \in \check{\mathcal{G}}^e_{j,s}$, $e \in E^*$, i.e., the corresponding average $\hat{u}_{j+1,s}$ is not available, because the cell $V_{j+1,s}$ is not contained in the adaptive grid. According to the construction of the adaptive grid this may happen only if there are no significant details corresponding to the underlying coarse grid cell $V_{j,k} \supset V_{j+1,s}$, otherwise $V_{j,k}$ would have been refined. In this case the details are zero. Therefore we introduce no error when applying (3.18). The exact representation would be $w_{j,r,e} - d_{j,r,e} = -\check{d}_{j,r,e}$ instead of $w_{j,r,e}$.

The reverse transformation is easier. For this purpose, we assume that there is an index $k \in I_j$ such that $T_{j,k} \subset J_{j,\varepsilon}$. According to one component of the full transformation (3.14) the *local inverse two–scale transformation* then reads

$$\hat{u}_{j+1,r} = \sum_{s \in \check{\mathcal{G}}^0_{j,r}} \check{g}^{j,0}_{s,r} \hat{u}_{j,s} + \sum_{e \in E^*} \sum_{s \in \check{\mathcal{G}}^e_{j,r}} \check{g}^{j,e}_{s,r} \check{d}_{j,s,e}, \quad r \in I^+_{j+1,k}, \tag{3.21}$$

$$w_{j,s,e} = \sum_{l \in \mathcal{L}^e_{j,s}} l^{j,e}_{l,s} \hat{u}_{j,l}, \quad s \in \overline{S}^e_{j,k}, \tag{3.22}$$

$$\check{d}_{j,s,e} = d_{j,s,e} - w_{j,s,e}, \quad s \in \overline{S}^e_{j,k}. \tag{3.23}$$

Again, the index sets $I^+_{j+1,k} \subset I_{j+1}$ and $\overline{S}^e_{j,k} \subset I_j$ are not yet determined. They will be derived when proving Lemma 4. There they are identified as

$$I^+_{j+1,k} = \check{\mathcal{M}}^0_{j,k}, \quad \overline{S}^e_{j,k} = \{k\}$$

according to the definitions (3.32) and (3.33).

3.4 Grading Parameter

In the previous section we have presented the local transformations according to (3.16) — (3.20) and (3.21) — (3.23), respectively. However it is not clear so far that these transformations can be efficiently realized. In particular, the number of floating point operations should be proportional to the number of cells in the adaptive grid and the number of significant details, respectively. Therefore it is prohibited to carry out the transformations on the full discretization levels which results in an effort proportional to the number of cells of the finest discretization. In order to develop an algorithm with an optimal complexity in the above sense the local transformations are implemented according to the following strategy:

- We proceed *levelwise* according to the pyramid scheme for the full transformation, see Fig. 2.6, i.e., all significant details and local averages belonging to the current level are computed before we proceed with the transformation on the next coarser (finer) level. In particular, local recursions should be avoided.
- We carry out *one* sweep through the refinement levels. This is only feasible if the set of significant details can be interpreted as a tree. Note, that this assumption is necessary for Algorithm 2 anyway.
- We compute only those components of the two–scale relations (3.13) and (3.14), respectively, which depend on significant details or local averages on higher scales.

In contrast to the transformation of the full discretization levels it is not obvious that the local two–scale relations (3.16) — (3.20) and (3.21) — (3.23) can be carried out, i.e., all data on the right hand side have already been computed and stored. In the following we will verify that the different steps of the local multiscale transformation are *feasible* in this sense. In a first step, we determine all indices corresponding to data that have to be made accessible before carrying out the different steps in the transformation. Here we distinguish between the encoding and the decoding transformation, see Lemma 3 and 4. From this we conclude the feasibility of the local transformations provided the grading parameter of the tree of significant details is chosen sufficiently large. For this purpose, we consider the following setting.

Assumption 1.

(1) the support $\mathcal{L}^e_{j,k}$ of the modification matrices $\mathsf{L}^e_{j,k}$ corresponding to the modified box wavelet is chosen such that

$$k \in \mathcal{L}^e_{j,k} \text{ or, equivalently, } k \in \mathcal{L}^{*,e}_{j,k} \qquad \forall\, k \in I_j, \tag{3.24}$$

(2) the tree of significant details $\mathcal{D}_{L,\varepsilon}$ is graded of degree q.

Then we derive sufficient conditions by which the local multiscale transformation can be verified to be feasible.

Lemma 3. *(Local multiscale transformation)*
Let Assumptions 1 be fulfilled. Let $k \in I_j$ such that $\check{\mathcal{M}}^0_{j,k} \subset I_{j+1,\varepsilon}$. If the condition

$$\pi_j(K^+_{j,k}) \subset \mathcal{N}^q_{j-1,\pi_j(k)} \text{ with } K^+_{j,k} = \bigcup_{e\in E^*} \bigcup_{s\in\mathcal{L}^{*,e}_{j,k}} \mathcal{L}^e_{j,s} \tag{3.25}$$

holds, then the local two–scale transformation (3.16) — (3.20) can be carried out.

Proof. First of all, we observe that the assumption $\check{\mathcal{M}}^0_{j,k} \subset I_{j+1,\varepsilon}$ is not restrictive. If we have already performed the local multiscale transformation for $j' = L, \ldots, j+1$, then this assumption also holds for level j. Hence, we conclude that the coarse–scale averages $\hat{u}_{j,r}$ can be computed according to (3.16) for all $r \in U^0_{j,k} := \bigcup_{r \in \check{\mathcal{M}}^0_{j,k}} \check{\mathcal{M}}^{*,0}_{j,r} = \{k\}$. Here we use the fact that

$$\check{\mathcal{M}}^{*,0}_{j,r} = \{k\} \qquad \forall\, r \in \check{\mathcal{M}}^0_{j,k}. \tag{3.26}$$

By assumption all fine–scale averages are available.

In addition to this, we have to compute the details $d_{j,r,e}$, $r \in U^e_{j,k}$, $e \in E^*$, depending on $\hat{u}_{j+1,s}$, $s \in \check{\mathcal{M}}^0_{j,k}$. According to (3.20) the set $U^e_{j,k}$ is determined by

$$U^e_{j,k} := \bigcup_{r \in \check{\mathcal{M}}^0_{j,k}} \mathcal{M}^{*,e}_{j,r} = \bigcup_{r \in \check{\mathcal{M}}^0_{j,k}} \bigcup_{l \in \check{\mathcal{M}}^{*,e}_{j,r}} \mathcal{L}^{*,e}_{j,l} = \bigcup_{r \in \check{\mathcal{M}}^0_{j,k}} \bigcup_{l \in \check{\mathcal{M}}^{*,0}_{j,r}} \mathcal{L}^{*,e}_{j,l} = \mathcal{L}^{*,e}_{j,k}, \tag{3.27}$$

where we incorporate (3.15) and (3.26).

The details are composed of two parts, namely, the details $\check{d}_{j,r,e}$ corresponding to the box wavelet and the modification terms $w_{j,r,e}$ due to the modification of the box wavelet. The details $\check{d}_{j,r,e}$, $r \in U^e_{j,k}$, are given by (3.19) where we access the averages $\hat{u}_{j+1,s}$, $s \in P^e_{j+1,k}$, with

$$P^e_{j+1,k} := \left(\bigcup_{r \in U^e_{j,k}} \check{\mathcal{M}}^e_{j,r} \right) \Big\backslash \check{\mathcal{M}}^0_{j,k} = \bigcup_{r \in \mathcal{L}^{*,e}_{j,k} \backslash \{k\}} \check{\mathcal{M}}^0_{j,r}. \tag{3.28}$$

Here we apply (3.15) and Assumption 3.24. Note, that the averages $\hat{u}_{j+1,s}$ have not necessarily been computed yet. However, we know for any $s \in P^e_{j+1,k}$ that all details in the two–scale representation of $\hat{u}_{j+1,s}$ are negligible and have been discarded by the thresholding procedure. Hence, we can reconstruct these averages only by coarse–scale averages according to (3.18).

For the computation of the details as well as for the reconstruction values, first of all the modification terms $w_{j,r,e}$ have to be determined for all indices $r \in S^e_{j,k}$ with

$$S^e_{j,k} := U^e_{j,k} \cup \bigcup_{s \in P^e_{j+1,k}} \check{\mathcal{G}}^e_{j,s} \tag{3.29}$$

according to (3.17). This requires the knowledge of the coarse–scale averages $\hat{u}_{j,l}$, $l \in K^+_{j,k}$, with

$$K^+_{j,k} := \bigcup_{e \in E^*} \left(\bigcup_{s \in P^e_{j+1,k}} \check{\mathcal{G}}^0_{j,s} \cup \bigcup_{s \in S^e_{j,k}} \mathcal{L}^e_{j,s} \right).$$

The representation of this index set becomes more simple if we take into account the definition of the involved index sets (3.28) and (3.29) as well as the fundamental relations for the supports of the box wavelet (3.15). First of all, we rewrite the set $S^e_{j,k}$. To this end, we observe that

$$\bigcup_{s\in P^e_{j+1,k}} \breve{\mathcal{G}}^e_{j,s} = \bigcup_{s\in\bigcup_{r\in\mathcal{L}^{*,e}_{j,k}\setminus\{k\}} \breve{\mathcal{M}}^0_{j,r}} \breve{\mathcal{M}}^{*,0}_{j,s} = \mathcal{L}^{*,e}_{j,k}\setminus\{k\} \qquad \forall e\in E,$$

where we make use of relation (3.26). This implies $S^e_{j,k} = \mathcal{L}^{*,e}_{j,k}$. From the definition of the set $K^+_{j,k}$ we now conclude together with assumption (3.24)

$$K^+_{j,k} = \bigcup_{e\in E^*}\left(\mathcal{L}^{*,e}_{j,k}\setminus\{k\}\cup\bigcup_{s\in\mathcal{L}^{*,e}_{j,k}}\mathcal{L}^e_{j,s}\right) = \bigcup_{e\in E^*}\bigcup_{s\in\mathcal{L}^{*,e}_{j,k}}\mathcal{L}^e_{j,s}.$$

According to Algorithm 2 the cells $V_{j,r}$, $r\in K^+_{j,k}$, are elements of the adaptive grid $\mathcal{G}_{L,\varepsilon}$ or will be further refined if and only if the details corresponding to the index set

$$\bigcup_{r\in K^+_{j,k}} T_{j-1,\pi_j(r)} \subset J_{j-1,\varepsilon} \tag{3.30}$$

are significant. On the other hand, we conclude from $k'\in I_{j+1,\varepsilon}$ that $T_{j,\pi_{j+1}(k')} = T_{j,k}\subset J_{j,\varepsilon}$. Then the grading implies

$$\bigcup_{r\in\mathcal{N}^q_{j-1,\pi_j(k)}} T_{j-1,r}\subset J_{j-1,\varepsilon}.$$

Hence, (3.30) is satisfied, if

$$\bigcup_{r\in K^+_{j,k}} T_{j-1,\pi_j(r)} \subset \bigcup_{r\in\mathcal{N}^q_{j-1,\pi_j(k)}} T_{j-1,r}.$$

This condition is equivalent to (3.25). □

In Fig. 3.9 the condition (3.25) is illustrated for the one–dimensional case according to Sect. 2.5.2. Here we assume that the stencil of the modified box wavelets is symmetric, i.e., $\mu = 0$ in (2.39). The cells corresponding to the index set $K^+_{j,k} = \{k-2\,s,\ldots,k+2\,s\}\subset I_j$ and the neighborhood $\mathcal{N}^q_{j-1,\pi_j(k)} = \mathcal{N}^q_{j-1,\lfloor k/2\rfloor} = \{\lfloor k/2\rfloor - q,\ldots,\lfloor k/2\rfloor + q\}\subset I_{j-1}$ are indicated by • and ∘, respectively. According to (3.5) and the uniform dyadic grid refinement, cf. Sects. 2.2 and 2.5.2, we obtain $\pi_j(K^+_{j,k}) = \{\lfloor k/2-s\rfloor,\ldots,\lfloor k/2+s\rfloor\}$. Obviously, condition (3.25) holds if $q\geq s$, cf. Corollary 4 in Sect. 3.8.

Note that condition (3.25) is sufficient to perform the local multiscale transformation. In particular, it is a constraint for the choice of the grading

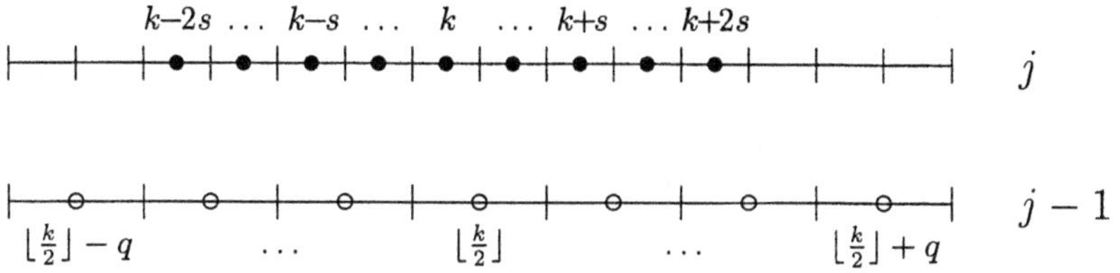

Fig. 3.9. Illustration of condition (3.25) and (3.31)

parameter q. In order to keep the transformation as efficient as possible we prefer q to be as small as possible.

For the local inverse multiscale transformation we obtain another constraint for the grading parameter.

Lemma 4. *(Local inverse multiscale transformation)*
Let Assumptions 1 be fulfilled. Let $k \in I_j$ such that $T_{j,k} \subset J_{j,\varepsilon}$. If the condition

$$\pi_j(\mathcal{L}^e_{j,k}) \subset \mathcal{N}^q_{j-1,\pi_j(k)} \tag{3.31}$$

holds, then the local inverse two-scale transformation (3.21) — (3.23) can be carried out.

Proof. First of all we determine all fine–scale averages $\hat{u}_{j+1,r}$, $r \in I^+_{j+1,k}$, influenced by the details $d_{j,k,e}$ which are assumed to be significant. According to (3.21) these are determined by

$$I^+_{j+1,k} := \bigcup_{e\in E^*} \check{\mathcal{G}}^{*,e}_{j,k} = \check{\mathcal{M}}^0_{j,k}, \tag{3.32}$$

where we incorporate the fundamental relations for the supports of the box wavelet (3.15). From (3.22) and (3.23) we conclude that the coarse–scale averages $\hat{u}_{j,s}$, $s \in \overline{S}^0_{j,k}$, and the details $\check{d}_{j,s,e}$, $s \in \overline{S}^e_{j,k}$, are accessed where

$$\overline{S}^e_{j,k} := \bigcup_{l\in I^+_{j+1,k}} \check{\mathcal{G}}^e_{j,l} = \bigcup_{\check{\mathcal{M}}^0_{j,k}} \mathcal{M}^{*,0}_{j,k} = \{k\}. \tag{3.33}$$

For the computation of the details according to (3.22) we first have to perform (3.23) where we additionally access the coarse–scale averages $\hat{u}_{j,s}$, $s \in \mathcal{L}^e_{j,k}$. Note, that the support $\mathcal{L}^e_{j,k}$ satisfies (3.24).

According to Algorithm 2 the cells $V_{j,r}$, $r \in \mathcal{L}^e_{j,k}$, are elements of the adaptive grid $\mathcal{G}_{L,\varepsilon}$ or will be refined further if and only if the details corresponding to the index set

$$\bigcup_{s \in \mathcal{L}^e_{j,k}} T_{j-1,\pi_j(s)} \subset J_{j-1,\varepsilon} \tag{3.34}$$

are significant. On the other hand we conclude from $T_{j,k} \subset J_{j,\varepsilon}$ and the grading that

$$\bigcup_{r \in \mathcal{N}^q_{j-1,\pi_j(k)}} T_{j-1,r} \subset J_{j-1,\varepsilon}.$$

Hence, (3.34) is satisfied, if

$$\bigcup_{s \in \mathcal{L}^e_{j,k}} T_{j-1,\pi_j(s)} \subset \bigcup_{r \in \mathcal{N}^q_{j-1,\pi_j(k)}} T_{j-1,r}.$$

This condition is equivalent to (3.31). □

Again, the resulting condition (3.31) is illustrated for the one–dimensional case according to Sect. 2.5.2. In Fig. 3.9, the cells corresponding to the index set $\mathcal{L}^e_{j,k} = \{k-s, \ldots, k+, s\} \subset I_j$ and the neighborhood $\mathcal{N}^q_{j-1,\pi_j(k)} = \{\lfloor k/2 \rfloor - q, \ldots, \lfloor k/2 \rfloor + q\} \subset I_{j-1}$ are indicated by • and ∘, respectively. According to (3.5) and the uniform dyadic grid refinement, cf. Sects. 2.2 and 2.5.2, we obtain $\pi_j(\mathcal{L}^e_{j,k}) = \{\lfloor (k-s)/2 \rfloor, \ldots, \lfloor (k+s)/2 \rfloor\}$. Obviously, condition (3.31) holds if $q \geq \lfloor s/2 \rfloor$ which is weaker than condition (3.25).

From Lemma 3 and 4 we now conclude the main result of this section.

Theorem 3. *(Local transformation)*
Let the Assumptions 1 be fulfilled. The local multiscale transformation as well as its inverse can be performed provided that condition (3.25) holds.

Proof. According to Lemma 3 and 4 the conditions (3.25) and (3.31) are sufficient for the feasibility of the local transformations. Since $\mathcal{L}^e_{j,k} \subset K^+_{j,k}$ condition (3.25) is stronger than (3.31). □

3.5 Locally Uniform Grids

The feasibility of the local multiscale transformations depends on the grading of the set of significant details $\mathcal{D}_{L,\varepsilon}$, see Theorem 3. In Sect. 3.2 we have already investigated the influence of the graded tree on the structure of the adaptive grid. Later, the evolution step of the averages will be performed on the adaptive grid. This requires the computation of the numerical fluxes at the interfaces of neighboring cells depending on a local stencil. For this purpose, we have to provide the averages corresponding to this stencil. In case of a structured flow solver as it is applied in the numerical experiments in Chapter 7, the adaptive grid should be locally uniform. This is to ensure that the reconstruction of averages on *one* local level does not involve all refinement levels but at most two. For this purpose, we have to specify what we mean by a locally uniform grid.

Definition 8. *(Locally uniform grid of degree p)*
An adaptive grid is said to be locally uniform of degree p, *if for any* $k \in I_{j,\varepsilon}$ *one of the three conditions holds for any* $l \in \mathcal{N}^p_{j,k}$:

(1) $l \in I_{j,\varepsilon}$ *or*

(2) $\check{\mathcal{M}}^0_{j,l} \subset I_{j+1,\varepsilon}$ *or*

(3) $T_{j-2,\pi_{j-1}(r)} \subset J_{j-2,\varepsilon} \quad \forall r \in \bigcup_{e\in E^*} \mathcal{L}^e_{j,\pi_j(l)}$.

If the grid is locally uniform of degree p, then we can access the averages $\hat{u}_{j,l}$, $l \in \mathcal{N}^p_{j,k}$, by

(1) either directly accessing the averages corresponding to the adaptive grid

(2) or locally applying the two–scale transformation, i.e.,

$$\hat{u}_{j,l} = \sum_{r\in\check{\mathcal{M}}^0_{j,l}} \check{m}^{j,0}_{r,l}\, \hat{u}_{j+1,r}.$$

(3) or locally applying the inverse two–scale transformation, i.e.,

$$\hat{u}_{j,l} = \sum_{r\in\check{\mathcal{G}}^0_{j-1,l}} \check{g}^{j-1,0}_{r,l}\, \hat{u}_{j-1,r} + \sum_{e\in E^*} \sum_{r\in\check{\mathcal{G}}^e_{j-1,l}} \check{g}^{j-1,e}_{r,l}\, d_{j-1,r,e},$$

$$\check{d}_{j-1,k,e} = d_{j-1,k,e} - w_{j-1,k,e}, \quad w_{j-1,k,e} = \sum_{m\in\mathcal{L}^e_{j-1,k}} l^{j-1,e}_{m,k}\, \hat{u}_{j-1,m}.$$

Again, if the tree of significant details is sufficiently graded, the adaptive grid is locally uniform.

Theorem 4. *(Graded grid ⇒ locally uniform grid)*
Assume that the Assumptions 1 are fulfilled. Additionally, the adaptive grid is assumed to be graded of degree $p \in \mathbb{N}$. *Whenever* $k \in I_j$ *such that there is* $k' \in \check{\mathcal{M}}^0_{j,k} \cap I_{j+1,\varepsilon}$, *then the adaptive grid is locally uniform of degree* p, *if the grading parameter* q *satisfies*

$$\pi_j\left(J^+_{j,k}\right) \subset \mathcal{N}^q_{j-1,\pi_j(k)}. \text{ with } J^+_{j,k} := \bigcup_{l\in\mathcal{N}^p_{j+1,k'}} \bigcup_{e\in E^*} \mathcal{L}^e_{j,\pi_{j+1}(l)} \qquad (3.35)$$

Proof. Let $k \in I_j$ such that $k' \in \check{\mathcal{M}}^0_{j,k} \cap I_{j+1,\varepsilon}$. Then we conclude from the definition of a graded grid that for all $l \in \mathcal{N}^p_{j+1,k'}$ one of the following conditions holds: (1) $l \in I_{j+1,\varepsilon}$ or (2) $l \notin I_{j+1,\varepsilon}$ and $\check{\mathcal{M}}^0_{j,l} \in I_{j+2,\varepsilon}$ or (3) $l \notin I_{j+1,\varepsilon}$ and $\pi_{j+1}(l) \in I_{j,\varepsilon}$. In case (1) there is nothing to do, since $\hat{u}_{j+1,l}$ is already available in the adaptive grid. If $l \notin I_{j+1,\varepsilon}$ the average $\hat{u}_{j+1,l}$ has to be computed by either coarsening or refining the grid. First we consider case (2). According to the local two–scale transformation we determine $\hat{u}_{j+1,l}$ by

$$\hat{u}_{j+1,l} = \sum_{r\in\check{\mathcal{M}}^0_{j+1,l}} \check{m}^{j+1,0}_{r,l}\, \hat{u}_{j+2,r}.$$

Finally, we have to check case (3). Here we have to apply the local inverse two-scale transformation in order to determine $\hat{u}_{j+1,l}$, i.e.,

$$\hat{u}_{j+1,l} = \sum_{r\in\check{\mathcal{G}}^0_{j,l}} \check{g}^{j,0}_{r,l}\,\hat{u}_{j,r} + \sum_{e\in E^*}\sum_{r\in\check{\mathcal{G}}^e_{j,l}} \check{g}^{j,e}_{r,l}\,d_{j,r,e},$$

$$\check{d}_{j,k',e} = d_{j,k',e} - w_{j,k',e}, \quad w_{j,k',e} = \sum_{m\in\mathcal{L}^e_{j,k'}} l^{j,e}_{m,k'}\,\hat{u}_{j,m}.$$

To carry these computation out, we obviously need the coarse–scale averages $\hat{u}_{j,m}$ for all indices m belonging to

$$\check{\mathcal{G}}^0_{j,l} \cup \bigcup_{e\in E^*}\bigcup_{r\in\check{\mathcal{G}}^e_{j,l}} \mathcal{L}^e_{j,r} = \check{\mathcal{M}}^{*,0}_{j,l} \cup \bigcup_{e\in E^*}\bigcup_{r\in\check{\mathcal{M}}^{*,0}_{j,l}} \mathcal{L}^e_{j,r} = \bigcup_{e\in E^*} \mathcal{L}^e_{j,\pi_{j+1}(l)}.$$

Here we incorporate the Assumptions 1. According to the refinement criterion these data are available if and only if

$$T_{j-1,\pi_j(r)} \subset J_{j-1,\varepsilon} \qquad \forall r \in \bigcup_{e\in E^*} \mathcal{L}^e_{j,\pi_{j+1}(l)}. \tag{3.36}$$

On the other hand, the refinement criterion and the choice of k and k', respectively, which are correlated by $k = \pi_{j+1}(k')$ imply that $T_{j,k} \subset J_{j,\varepsilon}$. Hence we conclude from the grading of the tree that

$$T_{j-1,r} \subset J_{j-1,\varepsilon} \qquad \forall r \in \mathcal{N}^q_{j-1,\pi_j(k)}. \tag{3.37}$$

In order to ensure (3.36) it suffices that

$$\bigcup_{l\in\mathcal{N}^p_{j+1,k'}} \;\bigcup_{r\in\bigcup_{e\in E^*}\mathcal{L}^e_{j,\pi_{j+1}(l)}} T_{j-1,\pi_j(r)} \subset \bigcup_{r\in\mathcal{N}^q_{j-1,\pi_j(k)}} T_{j-1,r}$$

holds. This condition results in a lower bound for the grading parameter q. It is fulfilled whenever (3.35) holds. □

In Fig. 3.10, the condition (3.35) is illustrated for the one–dimensional case according to Sect. 2.5.2. Here we assume that the stencil of the modified box wavelets is symmetric, i.e., $\mu = 0$ in (2.39). In particular, we conclude from the uniform dyadic grid hierarchy that $k' \in \check{\mathcal{M}}^0_{j,k} = \mathcal{M}_{j,k} = \{2k, 2k+1\}$ holds. According to (3.5) and the dyadic grid refinement, cf. Sects. 2.2 and 2.5.2, we determine the index sets of the neighborhoods $\mathcal{N}^p_{j+1,k'} = \{k' - p, \ldots, k' + p\} \subset I_{j+1}$ and $\mathcal{N}^q_{j-1,\pi_j(k)} = \{\lfloor k/2\rfloor - q, \ldots, \lfloor k/2\rfloor + q\} \subset I_{j-1}$ as well as the index set $J^+_{j,k} = \{k - s - \lfloor p/2\rfloor, \ldots, k + s + \lfloor p/2\rfloor\} \subset I_j$. These sets are indicated by • and ∘, respectively. From Fig. 3.10 it can be motivated that the condition (3.35) holds if $q \geq \lceil(\lceil p/2\rceil + s)/2\rceil$, cf. Corollary 5 in Sect. 3.8.

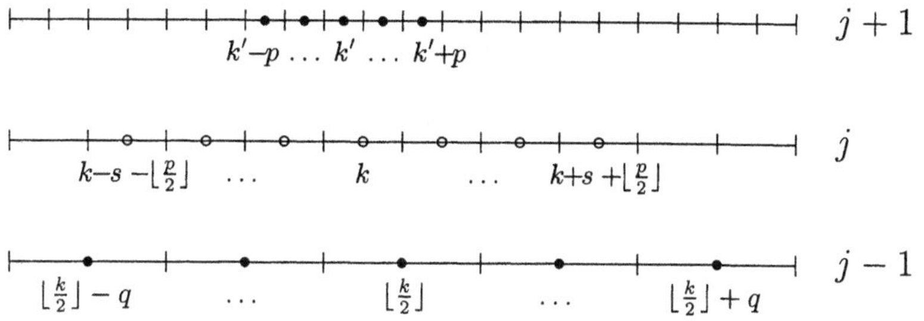

Fig. 3.10. Illustration of condition (3.35)

3.6 Algorithms: Encoding, Thresholding, Grading, Decoding

In the previous subsections we have analyzed to some extent the feasibility of the local transformations (3.16) — (3.20) and (3.21) — (3.23), respectively. In the following we will present the algorithms by which the local multiscale transformation and its inverse are realized in our computations. Note, that we comment on each step of an algorithm after having summarized the computational steps. The structure of the algorithms as well as the definition of the involved index sets is motivated by the proofs of Lemma 3 and 4, i.e., the proofs serve as a guideline for the computation. However, in the algorithm we proceed levelwise. Consequently, we do not carry out the local transformation for *one* single index but for all local averages and significant details that are influenced by local informations on the finer (coarser) level, respectively. In particular, the local multiscale transformation (3.16) — (3.20) is carried out for *all* $k \in I_j$ where the averages $\hat{u}_{j+1,r}$, $r \in \check{\mathcal{M}}^0_{j,k}$ have been computed in the previous step on level $j+1$ or the averages correspond to cells of the adaptive grid, respectively. For the local inverse transformation (3.21) — (3.23) we proceed similarly. Here we compute the averages $\hat{u}_{j+1,r}$, $r \in \check{\mathcal{M}}^0_{j,k}$, that are influenced by significant details $d_{j,k,e}$ for *all* $(k,e) \in \mathcal{J}_{j,\mathcal{E}}$.

First of all we consider the encoding algorithm. This algorithm needs the following input information: (i) the number of refinement levels L, (ii) the local index sets $I_{j,\mathcal{E}}$, $j = 0, \ldots, L$, representing the adaptive grid and (iii) the corresponding sequences of local averages $(\hat{u}_{j,k})_{k \in I_{j,\varepsilon}}$, $j = 0, \ldots, L$. Then the encoding algorithm provides the following output: (i) the local index sets of details $J_{j,\mathcal{E}}$, $j = 0, \ldots, L-1$, (ii) the corresponding sequences of details $(d_{j,k,e})_{(k,e) \in J_{j,\varepsilon}}$, $j = 0, \ldots, L-1$ and (iii) the averages $(\hat{u}_{0,k})_{k \in I_0}$ of the coarsest grid. We emphasize that the adaptive grid corresponding to $\mathcal{G}_{L,\mathcal{E}}$ may be determined by a set $\mathcal{D}_{L,\mathcal{E}}$ that does not correspond to the local averages $(\hat{u}_{j,k})_{(j,k) \in \mathcal{G}_{L,\varepsilon}}$ of the input data. This situation occurs indeed when performing an evolution step on the adaptive grid. Thus, the set $\mathcal{D}_{L,\mathcal{E}}$ of the output data is not necessarily graded and it might include non–significant details, because no thresholding is performed.

Algorithm 3. *(Local multiscale transformation)*
<u>for</u> $j = L-1$ <u>downto</u> *0* <u>do</u>

1. $U_j^0 := \bigcup_{r\in I_{j+1,\varepsilon}} \check{\mathcal{M}}_{j,r}^{*,0},\ U_j^e := \bigcup_{r\in I_{j+1,\varepsilon}} \mathcal{M}_{j,r}^{*,e},\ e\in E^*$

2. $\hat{u}_{j,k} := \sum_{r\in\check{\mathcal{M}}_{j,k}^0} \check{m}_{r,k}^{j,0}\,\hat{u}_{j+1,r},\ k\in U_j^0$

3. $P_{j+1}^e := \bigcup_{k\in U_j^e} \check{\mathcal{M}}_{j,k}^e \setminus I_{j+1,\varepsilon},\ e\in E^*,\ P_{j+1} := \bigcup_{e\in E^*} P_{j+1}^e$

4. $S_j^e := U_j^e \cup \bigcup_{l\in P_{j+1}^e} \check{\mathcal{G}}_{j,l}^e,\ e\in E^*$

5. $w_{j,k,e} := \sum_{r\in\mathcal{L}_{j,k}^e} l_{r,k}^{j,e}\hat{u}_{j,r},\ k\in S_j^e,\ e\in E^*$

6. $\hat{u}_{j+1,k} = \sum_{r\in\check{\mathcal{G}}_{j,k}^0} \check{g}_{r,k}^{j,0}\,\hat{u}_{j,r} - \sum_{e\in E^*}\sum_{r\in\check{\mathcal{G}}_{j,k}^e} \check{g}_{r,k}^{j,e}\,w_{j,e,r},\ k\in P_{j+1}$

7. <u>for</u> $e\in E^*$ <u>do</u>
  1. $\check{d}_{j,k,e} := \sum_{r\in\check{\mathcal{M}}_{j,k}^e} \check{m}_{r,k}^{j,e}\hat{u}_{j+1,r},\ k\in U_j^e$

  2. $d_{j,k,e} = w_{j,k,e} + \check{d}_{j,k,e},\ e\in E^*,\ k\in U_j^e$

8. <u>if</u> $k\notin U_j^0 \cup \bigcup_{(l,e)\in J_{j,\varepsilon}}\{l\}$ <u>then</u> *delete* $\check{m}_{r,k}^{j,0},\ r\in\check{\mathcal{M}}_{j,k}^0$

9. <u>for</u> $e\in E^*$ <u>do</u> <u>if</u> $k\notin S_j^e$ <u>then</u> *delete* $l_{r,k}^{j,e},\ r\in\mathcal{L}_{j,k}^e$

The local multiscale transformation is performed levelwise starting on the finest level. For each level we proceed as follows. In step 1 we first determine the indices of all coarse–scale averages and details which depend on the fine–scale averages of the adaptive grid. The sets U_j^e, $e\in E^*$, can be represented by

$$U_j^e = \bigcup_{r\in I_{j+1,\varepsilon}} \left(\check{\mathcal{M}}_{j,r}^{*,0} \cup \bigcup_{l\in\check{\mathcal{M}}_{j,r}^{*,e}} \mathcal{L}_{j,l}^{*,e} \right) = \bigcup_{r\in I_{j+1,\varepsilon}} \bigcup_{l\in\check{\mathcal{M}}_{j,r}^{*,e}} \mathcal{L}_{j,l}^{*,e} = \bigcup_{r\in I_{j+1,\varepsilon}} \mathcal{M}_{j,r}^{*,e},$$

if the relations $k\in\mathcal{L}_{j,k}^e$ and $k\in\mathcal{L}_{j,k}^{*,e}$, respectively, hold. Since the adaptive grid is assumed to be gap– and redundancy–free, see Definition 5, we know that all fine–scale averages required in the computation of the coarse–scale averages in step 2 are available. Otherwise an error occurs in the computation. Moreover, we have to determine the matrix entries of the kth row of $\check{\mathsf{M}}_{j,0}$ if not yet available. In step 3, we determine all indices for which an average on level $j+1$ will be needed. This reconstruction value is computed in step 6 and finally accessed in step 7.1. For the computation of these values we have to determine all indices for which the coarse–scale modification terms $w_{j,k,e}$, $e\in E^*$, have to be calculated, see step 4. The modification terms are then computed in step 5. Here we have to verify that all coarse–scale averages

$\hat{u}_{j,r}$ have already been computed. According to Theorem 3 this is guaranteed provided the details are sufficiently graded. Note, that we introduce in step 6 an error by the prediction value for $\hat{u}_{j+1,k}$, $k \in P_{j+1}$. According to (3.21) and (3.23) the error is given by

$$\sum_{e \in E^*} \sum_{r \in \check{\mathcal{G}}^e_{j,k}} \check{g}^{j,e}_{r,k} \left(\check{d}_{j,r,e} + w_{j,r,e}\right) = \sum_{e \in E^*} \sum_{r \in \check{\mathcal{G}}^e_{j,k}} \check{g}^{j,e}_{r,k} d_{j,r,e}.$$

By the assumption on the adaptive grid we know that the details $d_{j,r,e}$ are smaller than the threshold error ε_j; otherwise $V_{j+1,k}$ would be part of the adaptive grid and the average would have been computed, i.e., $k \in I_{j+1,\varepsilon}$. In step 7, we compute the details $d_{j,k,e}$. Since the computation is not only optimized with respect to computational time but also memory requirements, it is prohibited to compute and store *all* entries of the mask matrices. Instead, we have to manage those entries locally, i.e., only those columns are stored which are accessed in the local transformation process. This means that we have to compute the matrix entries of the kth column of $\check{\mathsf{M}}_{j,e}$, $\check{\mathsf{G}}_{j,e}$ and $\mathsf{L}^e_{j,k}$, if not yet available before performing the steps 2, 5, 6 and 7, respectively. In order to avoid an inflation of the matrices we have to remove all entries of $\check{\mathsf{M}}_{j,0}$ and $\mathsf{L}_{j,e}$, $e \in E^*$, corresponding to the kth column that have not been accessed in the transformation, see step 8 and 9.

Having performed the local multiscale transformation, the set of details $\mathcal{D}_{L,\varepsilon}$ is neither graded nor thresholded. Therefore, we have to present algorithms by which the non–significant details are discarded and the tree of details is graded. First of all, we outline the thresholding algorithm. Here we do not consider a scalar function $u \in L^1(\Omega, \mathsf{R})$ but a vector–valued function $u \in L^1(\Omega, \mathsf{R}^m)$ in view of applications to a system of conservation laws. For systems it is convenient to make the thresholding scale–invariant, i.e., the thresholding is invariant with respect to multiplying the averages by a constant factor. This becomes necessary because for systems of conservation laws no longer a maximum–minimum principle holds as in the scalar case. Hence, local extrema may develop and increase in time for each conservative quantity. If we do not introduce the scaling, then we have to choose threshold values ε_j for *each* component. For the scaling, we define the quantity $\|\mathbf{d}\|_{\infty,*}$ for a vector $\mathbf{d} \in \mathsf{R}^m$ by

$$\|\mathbf{d}\|_{\infty,*} := \|\mathbf{d}^*\|_\infty, \quad \mathbf{d}^* := \begin{cases} d_i & , \hat{u}_{\infty,i} < tol \\ \frac{d_i}{\hat{u}_{\infty,i}} & , \text{elsewhere} \end{cases}, \tag{3.38}$$

where *tol* is a machine depending tolerance to avoid division by 0. Here the components of the vector $\hat{\mathbf{u}}_\infty \in \mathsf{R}^m$ are defined by

$$\hat{u}_{\infty,i} := \max_{(j,k) \in \mathcal{G}_{L,\varepsilon}} |(\hat{u}_{j,k})_i|, \ i = 1, \ldots, m.$$

Obviously, the following input data are required: (i) the number of refinement levels L, (ii) the sequence of threshold values ε, (iii) the scaling parameter *tol*,

(iv) the local index sets $I_{j,\mathcal{E}}$, $j = 0,\dots,L$, representing the adaptive grid, (v) the corresponding sequences of local averages $(\hat{u}_{j,k})_{k\in I_{j,\varepsilon}}$, $j = 0,\dots,L$, (vi) the local index sets of details $J_{j,\mathcal{E}}$, $j = 0,\dots,L-1$ and (vii) the corresponding sequences of details $(d_{j,k,e})_{(k,e)\in J_{j,\varepsilon}}$, $j = 0,\dots,L-1$. Then the thresholding algorithm provides the following output: (i) the local index sets of *significant* details $J_{j,\mathcal{E}}$, $j = 0,\dots,L-1$ and (ii) the corresponding sequences of *significant* details $(d_{j,k,e})_{(k,e)\in J_{j,\varepsilon}}$, $j = 0,\dots,L-1$.

Algorithm 4. *(Thresholding)*
for $j = 0$ to *L-1* do

1. $\hat{u}_{\infty,i} := \max_{j=0,\dots,L} \max_{k\in I_{j,\mathcal{E}}} |(\hat{u}_{j,k})_i|$, $i = 1,\dots,m$

2. for $(k,e) \in J_{j,\mathcal{E}}$ do
  1. for $i = 1$ to m do

if $u_{\infty,i} < tol$ then $(d_{j,k,e})_i^* := (d_{j,k,e})_i$ else $(d_{j,k,e})_i^* := (d_{j,k,e})_i / u_{\infty,i}$

  2. $\|d_{j,k,e}\|_{\infty,*} := \|d^*_{j,k,e}\|_\infty$

3. for $(k,e) \in J_{j,\mathcal{E}}$ do

if $\max_{e'\in E^*} \|d_{j,k,e'}\|_{\infty,*} < \varepsilon_j$ then

$J_{j,\mathcal{E}} := J_{j,\mathcal{E}} \backslash (\{k\} \times E)$ *and delete* $d_{j,k,e'}$, $e' \in E$

In a preprocessing step we compute the maximum values $\hat{u}_\infty$ by means of the local averages $(\hat{u}_{j,k})_{(j,k)\in\mathcal{G}_{L,\varepsilon}}$. Then we determine the maximum of the details $d_{j,k,e}$ according to the scaling (3.38). If *all* details corresponding to the different wavelet types are smaller than the threshold value ε_j then these coefficients are non–significant and have to be discarded from $J_{j,\mathcal{E}}$ and $\mathcal{D}_{L,\mathcal{E}}$, respectively, and the corresponding values have to be removed from the memory. Here it is not necessarily required to proceed levelwise. Furthermore we like to remark that the thresholding can be alternatively performed within the local multiscale transformation, see Algorithm 3, where $\hat{u}_\infty$ has to be determined in a preprocessing step and the details computed in step 7.2 are only stored if $\|(d_{j,k,e})_{e\in E^*}\|_{\infty,*} \geq \varepsilon_j$. For the convenience of the reader, the local multiscale transformation and the thresholding are separated in order to emphasize where we have to distinguish between a scalar equation and a system of conservation laws. Note, that except for the thresholding all algorithms can be applied component–wise.

Before the inverse local multiscale transformation can be performed, the tree of details has to be graded, see Theorem 3. As input data we need: (i) the number of refinement levels L, (ii) the grading parameter q, (iii) the local index sets of significant details $J_{j,\mathcal{E}}$, $j = 0,\dots,L-1$ and (iv) the corresponding sequences of significant details $(d_{j,k,e})_{(k,e)\in J_{j,\varepsilon}}$, $j = 0,\dots,L-1$. Then the graded tree of significant details $\mathcal{D}_{L,\mathcal{E}}$ is determined as output of the following algorithm.

Algorithm 5. *(Grading)*
<u>for</u> $j = L - 1$ <u>downto</u> *1* <u>do</u>

1. $J^+_{j-1,\varepsilon} := \emptyset$

2. <u>for</u> $(k,e) \in J_{j,\varepsilon}$ <u>do</u> $J^+_{j-1,\varepsilon} := J^+_{j-1,\varepsilon} \cup (\mathcal{N}^q_{j-1,\pi_j(k)} \times E)$

3. <u>for</u> $(k,e) \in J^+_{j-1,\varepsilon} \backslash J_{j-1,\varepsilon}$ <u>do</u> $\mathbf{d}_{j-1,k,e} := \mathbf{0}$

4. $J_{j-1,\varepsilon} := J_{j-1,\varepsilon} \cup J^+_{j-1,\varepsilon}$

Here we have to proceed from the finest to the coarsest level. On each level we first have to initialize the grading set $J^+_{j-1,\varepsilon}$. Then we collect the neighborhoods $\mathcal{N}^q_{j-1,\pi_j(k)} \subset I_{j-1}$ of degree q for all significant details $(k,e) \in J_{j,\varepsilon}$ and set the corresponding details to be zero if not yet available. Finally, we have to add the grading set to the originally determined details on the lower level $j-1$.

Finally we consider the decoding algorithm. This algorithm needs the following input information: (i) the number of refinement levels L, (ii) the local index sets of significant details $J_{j,\varepsilon}$, $j = 0, \ldots, L-1$, (iii) the corresponding sequences of significant details $(d_{j,k,e})_{(k,e)\in J_{j,\varepsilon}}$, $j = 0, \ldots, L-1$ and (iv) the averages $(\hat{u}_{0,k})_{k\in I_0}$ of the coarsest grid. Note, that the set of significant details $\mathcal{D}_{L,\varepsilon}$ has to be graded in view of Theorem 3. Then the decoding algorithm provides the following output: (i) the local index sets $I_{j,\varepsilon}$, $j = 0, \ldots, L$, representing the adaptive grid and (ii) the corresponding sequences of local averages $(\hat{u}_{j,k})_{k\in I_{j,\varepsilon}}$, $j = 0, \ldots, L$.

Algorithm 6. *(Local inverse multiscale transformation)*
$I_0^+ := I_0$
<u>for</u> $j = 0$ <u>to</u> *L-1* <u>do</u>

1. $I^+_{j+1} := \bigcup_{(e,l)\in J_{j,\varepsilon}} \check{\mathcal{G}}^{*,e}_{j,l}$

2. $\overline{S}_j := \bigcup_{l\in I^+_{j+1}} \check{\mathcal{G}}^{*,0}_{j,l}$

3. $w_{j,k,e} := \sum_{r\in\mathcal{L}^e_{j,k}} l^{j,e}_{r,k}\, \hat{u}_{j,r},\ k \in \overline{S}_j,\ e \in E^*$

4. $\check{d}_{j,k,e} := d_{j,k,e} - w_{j,k,e},\ k \in \overline{S}_j,\ e \in E^*$

5. $\hat{u}_{j+1,k} = \sum_{r\in\check{\mathcal{G}}^0_{j,k}} \check{g}^{j,0}_{r,k}\, \hat{u}_{j,r} + \sum_{e\in E^*} \sum_{r\in\check{\mathcal{G}}^e_{j,k}} \check{g}^{j,e}_{r,k}\, \check{d}_{j,e,r},\ k \in I^+_{j+1}$

6. $I^-_j := \bigcup_{k\in I^+_{j+1}} \check{\mathcal{M}}^{*,0}_{j,k}$, $I_{j,\varepsilon} := I^+_j \backslash I^-_j$ *and delete* $\hat{u}_{j,k},\ k \in I^-_j$

7. <u>if</u> $k \notin I^+_{j+1} \cap \{k \,:\, \check{g}^{j,0}_{r,k}$ *exists*, $r \in \check{\mathcal{G}}^0_{j,k}\}$ <u>then</u> *delete* $\check{g}^{j,0}_{r,k},\ r \in \check{\mathcal{G}}^0_{j,k}$

In a preprocessing step we have to initialize the adaptive grid by *all* cells of the coarsest level. Then we perform the local inverse transformation in one sweep through the refinement levels starting from the coarsest resolution. On each level we proceed as follows. First we collect in step 1 all the fine–scale averages which are influenced by significant details. Then we determine all indices of coarse–scale averages involved in the computation of the modification term $w_{j,k,e}$, see step 2 and 3. Before we perform the summation in step 3 we have to check whether the matrix entries of the kth column of $\mathsf{L}_{j,e}$ are available. Otherwise we first have to determine these entries. In addition, we compute the details corresponding to the box wavelet according to step 4. Here we might access a non–significant detail $d_{j,k,e}$, i.e., $k \notin J_{j,\varepsilon}$. In this case we set $d_{j,k,e} := 0$. Finally, we calculate the fine–scale averages, see 5. Note, that an error occurs if the data $\hat{u}_{j,l}$ and $\check{d}_{j,l,e}$ are not available. In this case the set of significant details does not fit the assumptions, i.e., it is not graded. Moreover, we have to compute the matrix entries of the kth column of $\check{\mathsf{G}}_{j,e}$, $e \in E$, if not yet available. In order to avoid redundancies we remove all averages on the coarse scale which have been refined, see step 6. In particular, we clear the storage. To this end, we first determine all cells of level j that have been refined and remove them from the adaptive grid. This proceeding is analogical to Algorithm 2. Then we know all cells of the level j and the corresponding averages, i.e., $I_{j,\varepsilon}$ and $(\hat{u}_{j,k})_{k \in I_{j,\varepsilon}}$. Finally, we have to remove all columns of the matrices $\check{\mathsf{G}}_{j,0}$ which have not been used in the transformation step, see step 7.

3.7 Conservation Property

In Chapter 4, we will employ the local multiscale analysis for a finite volume method. For each time level, the adaptive grid is determined by means of the details corresponding to the multiscale decomposition of the averages at hand. Then we predict a new adaptive grid for the next time level. This requires the transformation of the averages from the old to the new grid where we access the sequence of *truncated* details. In view of the FVS the question arises whether the truncation process violates the conservation property of the reference FVS. However, the following lemma shows that the thresholding does not affect this property.

Lemma 5. *(Conservation property)*
Define the prediction values

$$\overline{u}_{j+1,r} := \sum_{l \in \mathcal{G}_{j,r}^0} \tilde{g}_{l,r}^{j,0} \, \hat{u}_{j,l}, \ r \in I_{j+1}. \tag{3.39}$$

Then the truncation process is locally mass preserving, i.e.,

$$\sum_{r \in \mathcal{M}_{j,k}^0} |V_{j+1,r}| \, \overline{u}_{j+1,r} = |V_{j,k}| \, \hat{u}_{j,k}. \tag{3.40}$$

and the global conservation property

$$\sum_{(j,k)\in\mathcal{G}_\varepsilon} |V_{j,k}|\,\hat{u}_{j,k} = \sum_{k\in I_0} |V_{0,k}|\,\hat{u}_{0,k} \tag{3.41}$$

holds.

Proof. First of all, we note that

$$\begin{aligned}\sum_{r\in\mathcal{M}^0_{j,k}} |V_{j+1,r}|\,\hat{u}_{j+1,r} &= |V_{j,k}| \sum_{r\in\mathcal{M}^0_{j,k}} \frac{|V_{j+1,r}|}{|V_{j,k}|}\,\hat{u}_{j+1,r} \\ &= |V_{j,k}| \sum_{r\in\mathcal{M}^0_{j,k}} \tilde{m}^{j,0}_{r,k}\,\hat{u}_{j+1,r} = |V_{j,k}|\,\hat{u}_{j,k}\end{aligned}$$

where we apply the representation (2.31) of the mask coefficients $\tilde{m}^{j,0}_{r,k}$ as well as the local two–scale transformation (3.16). Together with Definition 5 of the adaptive grid the global conservation property (3.41) follows.

In order to verify the local conservation property (3.40) we first consider the error introduced by the truncation, i.e.,

$$\begin{aligned}\sum_{r\in\mathcal{M}^0_{j,k}} |V_{j+1,r}|\,(\hat{u}_{j+1,r} - \overline{u}_{j+1,r}) &= |V_{j,k}| \sum_{r\in\mathcal{M}^0_{j,k}} \frac{|V_{j+1,r}|}{|V_{j,k}|}\,(\hat{u}_{j+1,r} - \overline{u}_{j+1,r}) \\ &= |V_{j,k}| \sum_{r\in\mathcal{M}^0_{j,k}} \tilde{m}^{j,0}_{r,k} \sum_{e\in E^*} \sum_{l\in\mathcal{G}^e_{j,r}} \tilde{g}^{j,e}_{l,r}\, d_{j,l,e},\end{aligned}$$

where we incorporate Definition (3.39) for the prediction values and the local inverse two–scale transformation (3.21). From the reversibility of the multiscale transformation we conclude that the error vanishes, since

$$\sum_{r\in\mathcal{M}^0_{j,k}} \tilde{m}^{j,0}_{r,k}\,\tilde{g}^{j,e}_{l,r} = 0$$

for all $e \in E^*$ and $r \in I_{j+1}$, $l \in J_{j,e}$, see also (2.13). Note, that the sets $\mathcal{M}^0_{j,k}$ and $\mathcal{G}^e_{j,r}$ can be replaced by the global index sets because the remaining mask coefficients are zero according to the definition of the supports. Finally, we obtain

$$\begin{aligned}&\sum_{r\in\mathcal{M}^0_{j,k}} |V_{j+1,r}|\,\overline{u}_{j+1,r} = \\ &\sum_{r\in\mathcal{M}^0_{j,k}} |V_{j+1,r}|\,\hat{u}_{j+1,r} + \sum_{r\in\mathcal{M}^0_{j,k}} |V_{j+1,r}|\,(\overline{u}_{j+1,r} - \hat{u}_{j+1,r})\end{aligned}$$

which immediately leads to (3.40) when applying the local multiscale transformation (3.16). □

From this lemma we conclude that all mass is stored in the averages but no mass is stored in the details. Hence the truncation of the details can not affect the conservation property of the adaptive FVS, since we perform no threshold on the coarse–scale averages[1].

3.8 Application to Curvilinear Grids

The numerical experiments presented in Chapter 7 have only been carried out for structured grids, in particular Cartesian and curvilinear grids. Therefore we summarize some computational aspects that arise when applying the above multiscale concept to structured grids. Here we consider all the steps that have to be performed when applying the concept to a concrete application which may serve as a concrete example.

Nested Grid Hierarchy. The starting point is a smooth function

$$\mathbf{x} : R := [0,1]^d \rightarrow \Omega$$

which maps the parameter domain R onto the computational domain Ω. The Jacobian is assumed to be regular, i.e., $\det(\partial\,\mathbf{x}(\xi)/\partial\,\xi) \neq 0,\ \xi \in R$. In our applications we represent the grid function by B–splines, see [BM00]. This admits control of good local grid properties, e.g., orthogonality and smoothness of the grid, and a consistent boundary representation by a small number of control points depending on the configuration at hand. Then a nested grid hierarchy according to Sect. 2.1 is defined by means of a sequence of nested uniform partitions of the parameter domain. To this end, we introduce the sets of multi–indices

$$I_j := \prod_{i=1}^{d} \{0, N_{j,i} - 1\} \subset \mathbb{N}_0^d, \qquad j = 0, \ldots, L,$$

with $N_{j,i} = 2\,N_{j-1,i}$ initialized by some $N_{0,i}$. Here the product denotes the Cartesian product, i.e., $\prod_{i=1}^d A_i := A_1 \times \cdots \times A_d$. Then the nested sequence of parameter partitions $\mathcal{R}_j := \{R_{j,\mathbf{k}}\}_{\mathbf{k}\in I_j}$, $j = 0, \ldots, L$, is determined by

$$R_{j,\mathbf{k}} := \prod_{i=1}^{d} [k_i\,h_{j,i}, (k_i + 1)\,h_{j,i}]$$

with $h_{j,i} := 1/N_{j,i} = h_{j-1,i}/2$. Finally the nested partitions of the computational domain are obtained by $V_{j,\mathbf{k}} := \mathbf{x}(R_{j,\mathbf{k}})$, see Fig. 3.11 for an illustration.

---

[1] In [CKMP01] the threshold process is also applied to the coarse–scale averages.

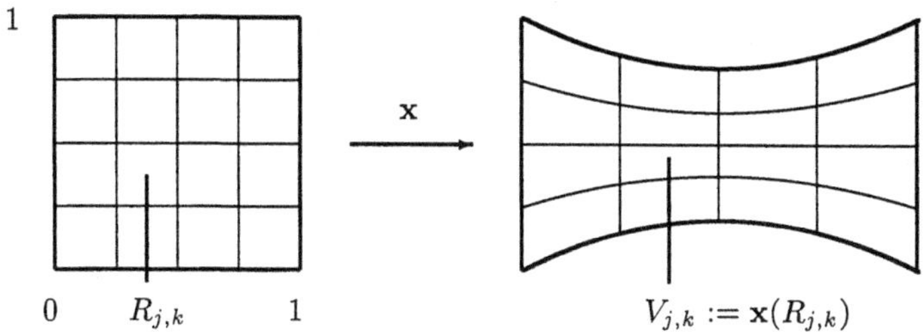

Fig. 3.11. Transformation from parameter to computational domain

Obviously, the refinement relation for the cells holds with

$$\mathcal{M}_{j,\mathbf{k}} = \{2\mathbf{k} + \mathbf{i}\ ;\ \mathbf{i} \in E := \{0,1\}^d\},$$

i.e., each cell is subdivided into $M_r = 2^d$ subcells. In this case, the sequence $\{\mathcal{G}_j\}_{j=0,\ldots,L}$ is referred to as a *dyadic* grid hierarchy. We emphasize that the set E can be identified with the index set $E = \{0, \ldots, M_r - 1\}$ by which the different types of box wavelets are indicated in Sect. 2.3. As a special application of this curvilinear setting we have already considered the Cartesian grid hierarchy, cf. Sect. 2.3.1.

Construction of Box Wavelets. In order to construct the box wavelets according to Sect. 2.3 we have to determine appropriate matrices $\mathsf{A}_{j,\mathbf{k}}$. Stability of the multiscale transformation is ensured when the rows of the matrix form an *orthonormal* system whereas the *linear independence* of the rows already provides a valid complement basis. Note, that in case of a curvilinear grid the matrices $\mathsf{A}_{j,\mathbf{k}}$ in general differ for all $(j, \mathbf{k})$. However, they coincide provided that the grid refinement is *uniform*, i.e.,

$$|V_{j+1,\mathbf{r}}|/|V_{j,\mathbf{k}}| = c_{j,\mathbf{k}} = 2^{-d} \quad \forall \mathbf{r} \in \mathcal{M}^0_{j,\mathbf{k}}, \tag{3.42}$$

cf. Sect. 2.3.1. In the sequel, we are therefore interested in a construction of matrices $\mathsf{A}_{j,\mathbf{k}}$ that leads to an *efficient* computation. To this end, we assume that the matrices are enumerated by the index set of multi–indices, i.e.,

$$\mathsf{A}_{j,\mathbf{k}} = \{a^{j,\mathbf{k}}_{\mathbf{e},\mathbf{r}}\}_{\mathbf{e}\in E,\mathbf{r}\in\mathcal{M}^0_{j,\mathbf{k}}} = \{a^{j,\mathbf{k}}_{\mathbf{e},2\mathbf{k}+\mathbf{i}}\}_{\mathbf{e},\mathbf{i}\in E}.$$

Then the starting point for the construction is the matrix $\mathsf{A} := \{(-1)^{\mathbf{e}\cdot\mathbf{i}}\}_{\mathbf{e},\mathbf{i}\in E}$ which is symmetric and, up to a factor, orthogonal. The inverse is given by $\mathsf{A}^{-1} = 2^{-d}\,\mathsf{A}$, see [Got98], Lemma 3.2, p. 77. Note, that A and A^{-1} are independent of j and $\mathbf{k}$. We now define the matrix $\mathsf{A}_{j,\mathbf{k}} := \mathsf{A}\,\boldsymbol{\Lambda}_{j,\mathbf{k}}$ by means of the coefficients

$$a^{j,\mathbf{k}}_{\mathbf{e},2\mathbf{k}+\mathbf{i}} = (-1)^{\mathbf{e}\cdot\mathbf{i}} \left(\frac{|V_{j+1,2\mathbf{k}+\mathbf{i}}|}{|V_{j,\mathbf{k}}|}\right)^{1/2} = (-1)^{\mathbf{e}\cdot\mathbf{i}} a^{j,\mathbf{k}}_{\mathbf{0},2\mathbf{k}+\mathbf{i}}. \tag{3.43}$$

The inverse of this matrix exists and is determined by $\mathsf{A}_{j,\mathbf{k}}^{-1} = 2^{-d}\,\mathbf{\Lambda}_{j,\mathbf{k}}^{-1}\,\mathsf{A} = \mathbf{\Lambda}_{j,\mathbf{k}}^{-1}\,\mathsf{A}^{-1}$ with the coefficients

$$b_{2\mathbf{k}+\mathbf{i},\mathbf{e}}^{j,\mathbf{k}} = 2^{-d}\,(-1)^{\mathbf{e}\cdot\mathbf{i}}\left(\frac{|V_{j,\mathbf{k}}|}{|V_{j+1,2\mathbf{k}+\mathbf{i}}|}\right)^{1/2} = 2^{-d}\,(-1)^{\mathbf{e}\cdot\mathbf{i}}(a_{\mathbf{0},2\mathbf{k}+\mathbf{i}}^{j,\mathbf{k}})^{-1}. \quad (3.44)$$

Note, that the inverse matrix is *not* determined by $\mathsf{A}_{j,\mathbf{k}}^{T}$, since the matrix $\mathsf{A}_{j,\mathbf{k}}$ is in general not orthogonal due to the relation

$$\overline{\mathbf{\Lambda}}_{j,\mathbf{k}} := \mathsf{A}_{j,\mathbf{k}}\,\mathsf{A}_{j,\mathbf{k}}^{T} = \mathsf{A}\,\mathbf{\Lambda}_{j,\mathbf{k}}^{2}\mathsf{A} = 2^{d}\,\mathsf{A}^{-1}\,\mathbf{\Lambda}_{j,\mathbf{k}}^{2}\mathsf{A}.$$

This only becomes the identity matrix, if the grid refinement satisfies (3.42), e.g. Cartesian grids. For this special configuration one obtains $\mathbf{\Lambda}_{j,\mathbf{k}}^{2} = 2^{-d}\,\mathsf{I}$, cf. Sect. 2.3.1. However, in the case of a curvilinear grid based on a smooth grid function the grid refinement is only locally approximately equidistant, in particular, for higher refinement levels j, i.e., $\mathbf{\Lambda}_{j,\mathbf{k}}^{2} \sim 2^{-d}\,\mathsf{I}$.

We emphasize that in case of a non–orthogonal matrix, the box wavelets and their L^{∞}–normalized counterparts do not form an biorthogonal system. Hence we can not apply Corollary 1, i.e., $\check{\mathsf{M}}_{j,1}$ is not necessarily a *stable* completion of $\check{\mathsf{M}}_{j,0}$ if the matrix $\mathsf{A}_{j,\mathbf{k}}$ is not orthogonal. Nevertheless, the system is supposed to be stable in the sense of Sect. 2.4 due to the above local perturbation argument. This is justified by the numerical results. In spite of the missing orthogonality property, the resulting box wavelets form a basis, since the matrices $\mathsf{A}_{j,\mathbf{k}}$ are invertible.

Finally we present the mask coefficients of the corresponding refinement equations. For the box wavelet we obtain for $\mathbf{k},\mathbf{l} \in I_j, e \in E$

$$\check{m}_{2\mathbf{l}+\mathbf{i},\mathbf{k}}^{j,\mathbf{e}} = \frac{|V_{j+1,2\mathbf{k}+\mathbf{i}}|}{|V_{j,\mathbf{k}}|}\,(-1)^{\mathbf{e}\cdot\mathbf{i}}\,\delta_{\mathbf{l},\mathbf{k}}, \quad \check{g}_{\mathbf{k},2\mathbf{l}+\mathbf{i}}^{j,\mathbf{e}} = 2^{-d}\,\frac{|V_{j,\mathbf{k}}|}{|V_{j+1,2\mathbf{k}+\mathbf{i}}|}\,(-1)^{\mathbf{e}\cdot\mathbf{i}}\,\delta_{\mathbf{l},\mathbf{k}},$$

where we replace $a_{e,r}^{j,k}$ and $b_{r,e}^{j,k}$ in (2.31) by $a_{\mathbf{e},2\mathbf{k}+\mathbf{i}}^{j,\mathbf{k}}$ and $b_{2\mathbf{k}+\mathbf{i},\mathbf{e}}^{j,\mathbf{k}}$ according to (3.43) and (3.44).

Modification of Box Wavelets. In order to increase the number of vanishing moments of the box wavelet we apply Algorithm 1. To this end, we have to fix the stencils $\mathcal{L}_{j,\mathbf{k}}^{\mathbf{e}} \subset I_j$. Here we choose only one stencil for all $\mathbf{e} \in E$. To obtain a small support we choose a local stencil that is located around the cell $V_{j,\mathbf{k}}$ where the box wavelets $\check{\psi}_{j,\mathbf{k},\mathbf{e}}$, $\mathbf{e} \in E$, correspond to, i.e.,

$$\mathcal{L}_{j,\mathbf{k}}^{s} := \prod_{i=1}^{d} \mathcal{L}_{j,k_i}^{s} = \prod_{i=1}^{d} \{l_{j,k_i}, \ldots, l_{j,k_i} + 2s\}, \qquad s \in \mathbb{N}_0, \quad (3.45)$$

with

$$l_{j,k} = \begin{cases} 0 & , 0 \le k_i \le s-1 \\ k_i - s & , s \le k_i \le N_{j,i} - 1 - s \\ N_{j,i} - 1 - 2s & , N_{j,i} - s \le k_i \le N_{j,i} - 1 \end{cases}.$$

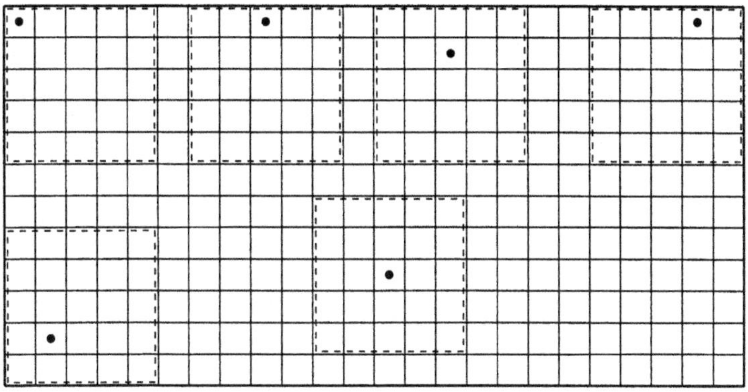

Fig. 3.12. Illustration of the sets $\mathcal{L}^s_{j,\mathbf{k}}$ for $s = 2$ and $d = 2$

Here the stencil width is $2s + 1 \leq N_{j,i}$ in each direction. Inside the computational domain Ω a central stencil is chosen, i.e., s cells at each side of $V_{j,\mathbf{k}}$. Close to the boundary we shift the central stencil such that the support is completely contained in Ω. An illustration of the sets in the bivariate case $d = 2$ is shown in Fig. 3.12. Here the dots indicate the position $\mathbf{k}$ and the dotted lines border the stencils. For the one–dimensional case, this setting has already been considered in Sect. 2.5.2. In order to minimize the stencil width with respect to a given number of vanishing moments M_c we choose

$$s := \min \{r \in \mathbb{N}_0 \; ; \; (2r + 1)^d > M_c\}. \tag{3.46}$$

Finally, we present the mask coefficients of the modified box wavelet

$$\tilde{m}^{j,\mathbf{e}}_{2\mathbf{l}+\mathbf{i},\mathbf{k}} = \frac{|V_{j+1,2\mathbf{l}+\mathbf{i}}|}{|V_{j,\mathbf{l}}|} \, (l^{j,\mathbf{e}}_{\mathbf{l},\mathbf{k}} + (-1)^{\mathbf{e}\cdot\mathbf{i}} \, \delta_{\mathbf{k},\mathbf{l}}), \quad \mathbf{e} \in E,$$

$$\tilde{g}^{j,\mathbf{0}}_{\mathbf{k},2\mathbf{l}+\mathbf{i}} = 2^{-d} \, \frac{|V_{j,\mathbf{l}}|}{|V_{j+1,2\mathbf{l}+\mathbf{i}}|} \left(\delta_{\mathbf{k},\mathbf{l}} - \sum_{\mathbf{e} \in E^*} (-1)^{\mathbf{e}\cdot\mathbf{i}} \, l^{j,\mathbf{e}}_{\mathbf{k},\mathbf{l}} \right),$$

where we have applied (3.43) and (3.44) in (2.38).

Note that for a uniform Cartesian grid hierarchy, cf. Sect. 2.3.1, appropriate modified box wavelets with higher vanishing moments can alternatively be constructed according to (2.18) using tensor products of the univariate modified box wavelets determined in Sect. 2.5.2.

Supports of Box Wavelets and Their Modification. For the representation of the local multiscale setting, see Sect. 3, we introduced the supports of the columns and rows corresponding to the mask matrices. In the case of the curvilinear setting these can explicitly be determined. First of all, we observe that all supports can be represented as Cartesian products, i.e.,

– $\mathcal{M}^{\mathbf{e}}_{j,\mathbf{k}} = \prod_{\mu=1}^d \mathcal{M}^{e_\mu}_{j,k_\mu}$, $\mathcal{G}^{\mathbf{e}}_{j,\mathbf{k}} = \prod_{\mu=1}^d \mathcal{G}^{e_\mu}_{j,k_\mu}$, $\mathbf{e} \in E$,

- $\mathcal{M}^{*,\mathbf{e}}_{j,\mathbf{k}} = \prod_{\mu=1}^{d} \mathcal{M}^{*,e_\mu}_{j,k_\mu}$, $\mathcal{G}^{*,\mathbf{e}}_{j,\mathbf{k}} = \prod_{\mu=1}^{d} \mathcal{G}^{*,e_\mu}_{j,k_\mu}$, $\mathbf{e} \in E$,
- $\mathcal{L}^{s}_{j,\mathbf{k}} = \prod_{\mu=1}^{d} \mathcal{L}^{s}_{j,k_\mu}$, $\mathcal{L}^{s}_{j,k_\mu} := \{l_{j,k_\mu}, \ldots, l_{j,k_\mu} + 2s\}$,
- $\mathcal{L}^{*,s}_{j,\mathbf{k}} = \prod_{\mu=1}^{d} \mathcal{L}^{*,s}_{j,k_\mu}$, $\mathcal{L}^{*,s}_{j,k_\mu} := \{\bar{l}^{0}_{j,k_\mu}, \ldots, \bar{l}^{1}_{j,k_\mu}\}$.

Here the numbers $\bar{l}^{0}_{j,k_\mu}$, $\bar{l}^{1}_{j,k_\mu}$, $\mu = 1, \ldots, d$, are determined by Lemma 6 and l_{j,k_μ} by the stencil $\mathcal{L}^{s}_{j,\mathbf{k}}$ according to the modification of the box wavelets. In particular, for the box function and the box wavelets the supports are given by

- $\check{\mathcal{M}}^{\mathbf{e}}_{j,\mathbf{k}} = \check{\mathcal{G}}^{*,\mathbf{e}}_{j,\mathbf{k}} = \{2\mathbf{k} + \mathbf{i} \; ; \; \mathbf{i} \in E\}$, $\mathbf{k} \in I_j$, $\mathbf{e} \in E$,
- $\check{\mathcal{M}}^{*,\mathbf{e}}_{j,\mathbf{k}} = \check{\mathcal{G}}^{\mathbf{e}}_{j,\mathbf{k}} = \{\lfloor \mathbf{k}/2 \rfloor\}$, $\mathbf{k} \in I_{j+1}$, $\mathbf{e} \in E_j$,

where $\lfloor k/2 \rfloor = l$ is the largest integer l not larger than $\lfloor k/2 \rfloor$. Then we obtain the supports of the modified box wavelets

- $\mathcal{M}^{0}_{j,\mathbf{k}} = \check{\mathcal{M}}^{0}_{j,\mathbf{k}} = \{2\mathbf{k} + \mathbf{i} \; ; \; \mathbf{i} \in E\}$, $\mathbf{k} \in I_j$,
- $\mathcal{M}^{\mathbf{e}}_{j,\mathbf{k}} = \{\mathbf{r} \in I_{j+1}; \lfloor \mathbf{r}/2 \rfloor \in \mathcal{L}^{s}_{j,\mathbf{k}}\}$
  $= \prod_{\mu=1}^{d} \{2l_{j,k_\mu}, \ldots, 2l_{j,k_\mu} + 4s + 1\}$, $\mathbf{k} \in I_j$, $\mathbf{e} \in E^*$,
- $\mathcal{G}^{0}_{j,\mathbf{k}} = \mathcal{L}^{s}_{j,\lfloor \mathbf{k}/2 \rfloor}$, $\mathbf{k} \in I_{j+1}$,
- $\mathcal{G}^{\mathbf{e}}_{j,\mathbf{k}} = \check{\mathcal{G}}^{\mathbf{e}}_{j,\mathbf{k}} = \{\lfloor \mathbf{k}/2 \rfloor\}$, $\mathbf{k} \in I_{j+1}$, $\mathbf{e} \in E^*$,
- $\mathcal{M}^{*,0}_{j,\mathbf{k}} = \check{\mathcal{M}}^{*,0}_{j,\mathbf{k}} = \check{\mathcal{G}}^{0}_{j,\mathbf{k}} = \{\lfloor \mathbf{k}/2 \rfloor\}$, $\mathbf{k} \in I_{j+1}$,
- $\mathcal{M}^{*,\mathbf{e}}_{j,\mathbf{k}} = \mathcal{L}^{*,s}_{j,\lfloor \mathbf{k}/2 \rfloor}$, $\mathbf{k} \in I_{j+1}$, $\mathbf{e} \in E^*$,
- $\mathcal{G}^{*,0}_{j,\mathbf{k}} = \{\mathbf{r} \; ; \; \lfloor \mathbf{r}/2 \rfloor \in \mathcal{L}^{*,s}_{j,\mathbf{k}}\} = \prod_{\mu=1}^{d} \{2\bar{l}^{0}_{j,k_\mu}, \ldots, 2\bar{l}^{1}_{j,k_\mu} + 1\}$, $\mathbf{k} \in I_j$,
- $\mathcal{G}^{*,\mathbf{e}}_{j,\mathbf{k}} = \check{\mathcal{G}}^{*,\mathbf{e}}_{j,\mathbf{k}} = \check{\mathcal{M}}^{\mathbf{e}}_{j,\mathbf{k}} = \{2\mathbf{k} + \mathbf{i} \; ; \; \mathbf{i} \in E\}$, $\mathbf{k} \in I_j$, $\mathbf{e} \in E^*$.

The construction of the support $\mathcal{L}^{s}_{j,\mathbf{k}}$ is more complicated because there exist three different types of matrix structures sketched below for the one–dimensional case ($N_j = 12,\ 10,\ 8$, respectively, and $s = 2$).

1) $2s + 1 < N_j - 1 - 2s$:

$$\begin{pmatrix}
+ & * & * & * & * & & & & & & & \\
* & + & * & * & * & & & & & & & \\
* & * & + & * & * & & & & & & & \\
 & * & * & + & * & * & & & & & & \\
 & & * & * & + & * & * & & & & & \\
 & & & * & * & + & * & * & & & & \\
 & & & & * & * & + & * & * & & & \\
 & & & & & * & * & + & * & * & & \\
 & & & & & & * & * & + & * & * & \\
 & & & & & & & * & * & + & * & * \\
 & & & & & & & * & * & * & + & * \\
 & & & & & & & * & * & * & * & +
\end{pmatrix}$$

2) $2s + 1 = N_j - 1 - 2s$:

$$\begin{pmatrix}
+ & * & * & * & * & & & & & \\
* & + & * & * & * & & & & & \\
* & * & + & * & * & & & & & \\
& * & * & + & * & * & & & & \\
& & * & * & + & * & * & & & \\
& & & * & * & + & * & * & & \\
& & & & * & * & + & * & * & \\
& & & & & * & * & + & * & * \\
& & & & & * & * & * & + & * \\
& & & & & * & * & * & * & +
\end{pmatrix}$$

3) $2s+1 > N_j - 1 - 2s$:

$$\begin{pmatrix}
+ & * & * & * & * & & & \\
* & + & * & * & * & & & \\
* & * & + & * & * & & & \\
& * & * & + & * & * & & \\
& & * & * & + & * & * & \\
& & & * & * & + & * & * \\
& & & * & * & * & + & * \\
& & & * & * & * & * & +
\end{pmatrix}$$

Employing these different matrix structures the supports of $\mathcal{L}^{*,s}_{j,\mathbf{k}}$ are determined by the following lemma.

Lemma 6. *(Construction of $\mathcal{L}^{*,s}_{j,\mathbf{k}}$)*
Let $N_j \geq 2s+1$. Then the support of the k–th column of $\mathsf{L}^s_{j,\mathbf{k}}$ is determined by the upper and lower bounds

$$\bar{l}^0_{j,k_\mu} = \begin{cases} 0 & , 0 \leq k_\mu \leq 2s \\ k_\mu - s & , 2s+1 \leq k_\mu \leq N_{j,\mu} - 1 \end{cases},$$

$$\bar{l}^1_{j,k_\mu} = \begin{cases} k_\mu + s & , 0 \leq k_\mu \leq N_{j,\mu} - 2s - 2 \\ N_{j,\mu} - 1 & , N_{j,\mu} - 2s - 1 \leq k_\mu \leq N_{j,\mu} - 1 \end{cases}.$$

Proof. Since the support $\mathcal{L}^{*,s}_{j,\mathbf{k}}$ is a Cartesian product, it suffices to consider only one component. Therefore we assume $d = 1$ and omit the subindex μ for the direction. Then we consider the following decomposition of the support $\mathcal{L}^{*,s}_{j,k} = \mathcal{M}_1 \cup \mathcal{M}_2 \cup \mathcal{M}_3$ with

$$\mathcal{M}_1 := \{l \in I_j \; ; \; 0 \leq l \leq s-1, \; k \in \mathcal{L}^s_{j,l}\},$$

$$\mathcal{M}_2 := \{l \in I_j \; ; \; s \leq l \leq N_j - s - 1, \; k \in \mathcal{L}^s_{j,l}\},$$

$$\mathcal{M}_3 := \{l \in I_j \; ; \; N_j - s \leq l \leq N_j - 1, \; k \in \mathcal{L}^s_{j,l}\}.$$

From this we conclude

$$\mathcal{M}_1 = \begin{cases} \{0, \ldots, s-1\} & , 0 \leq k \leq 2s \\ \emptyset & , elsewhere \end{cases},$$

$$\mathcal{M}_2 = \{l \in I_j \; ; \; \max\{s, k-s\} \leq l \leq \min\{N_j - s - 1, k+s\}\},$$

$$\mathcal{M}_3 = \begin{cases} \{N_j - s, \ldots, N_j - 1\} & , N_j - s - 2 \leq k \leq N_j - 1 \\ \emptyset & , elsewhere \end{cases}.$$

Then we have to distinguish six cases according to the three different matrix structures of $\mathsf{L}^s_{j,k}$.

1a) $N_j \geq 4s+2$ and $0 \leq k \leq 2s$:

$\mathcal{M}_1 = \{0,\ldots,s-1\}$, $\mathcal{M}_2 = \{s,\ldots,k+s\}$, $\mathcal{M}_3 = \emptyset$,

1b) $N_j \geq 4s+2$ and $2s+1 \leq k \leq N_j - 2s - 2$:

$\mathcal{M}_1 = \emptyset$, $\mathcal{M}_2 = \{k-s,\ldots,k+s\}$, $\mathcal{M}_3 = \emptyset$,

1c) $N_j \geq 4s+2$ and $2s+1 \leq k \leq N_j - 2s - 2$:

$\mathcal{M}_1 = \emptyset$, $\mathcal{M}_2 = \{k-s,\ldots,k+s\}$, $\mathcal{M}_3 = \emptyset$,

2a) $N_j \leq 4s+2$ and $0 \leq k \leq N_j - 2s - 2$:

$\mathcal{M}_1 = \{0,\ldots,s-1\}$, $\mathcal{M}_2 = \{s,\ldots,k+s\}$, $\mathcal{M}_3 = \emptyset$,

2b) $N_j \leq 4s+2$ and $\leq N_j - 2s - 1 \leq k \leq 2s$:

$\mathcal{M}_1 = \{0,\ldots,s-1\}$, $\mathcal{M}_2 = \{s,\ldots,N_j-s-1\}$, $\mathcal{M}_3 = \{0,\ldots,N_j-1\}$,

2c) $N_j \leq 4s+2$ and $2s+1 \leq k \leq N_j - 1$:

$\mathcal{M}_1 = \emptyset$, $\mathcal{M}_2 = \{k-s,\ldots,N_j-1-s\}$, $\mathcal{M}_3 = \{N_j-s,\ldots,N_j-1-s\}$.

This verifies the assertion. □

Neighborhoods. For the grading of the details as well as the definition of locally uniform grids, see Definition 7 and 8, we need the neighborhoods $\mathcal{N}^q_{j,\mathbf{k}} \subset I_j$ of degree q. In case of structured grids as they are considered here these are determined by

$$\mathcal{N}^q_{j,\mathbf{k}} = \prod_{i=1}^{d} \{\max(k_i - q, 0), \ldots, \min(N_{j,i} - 1, k_i + q)\}.$$

Note, that in case of an unbounded domain we do not have to take the maximum and minimum values into account but use instead the set of integers $\{k_i - q, \ldots, k_i + q\}$.

Grading Parameter for Modified Box Wavelets. In view of the feasibility of the local multiscale transformation and its inverse we have to verify Theorem 3. The result is motivated by the illustration in Fig. 3.9 for the one–dimensional case according to Sect. 2.5.2. Here we assume that the stencil of the modified box wavelets is symmetric, i.e., $\mu = 0$ in (2.39).

Corollary 4. *(Verification of local transformation)*
In the curvilinear case the local multiscale transformation as well as its inverse are feasible, if the grading parameter satisfies

$$q \geq s.$$

Proof. In order to verify the assertion we apply Theorem 3. Since we consider the curvilinear case, all supports can be interpreted as Cartesian products. Hence, it suffices to consider the one–dimensional case in the subsequent proof.

In order to choose the appropriate grading parameter q we first have to determine the index set $K^+_{j,k}$ according to (3.25). For the special choice $\mathcal{L}^e_{j,k} \equiv \mathcal{L}^s_{j,k}$, $\forall e \in E^* = \{1\}$ this set can be represented as

$$K^+_{j,k} = \bigcup_{r \in \mathcal{L}^{*,s}_{j,k}} \mathcal{L}^s_{j,r} = \bigcup_{i=\bar{l}^0_{j,k}}^{\bar{l}^1_{j,k}} \{l_{j,i}, \ldots, l_{j,i} + 2s\} = \{\check{l}^0_{j,k}, \ldots, \check{l}^1_{j,k}\}$$

where $l_{j,i}$ and $\bar{l}^0_{j,k}, \bar{l}^1_{j,k}$ are determined according to the previous subsections.

We now determine the lower and upper bounds of the index sets depending on the level j and the position k. To this end, we distinguish six cases corresponding to N_j and k.

1.) $N_j \le 4s + 1$:

a) $K^+_{j,k} = \{0, \ldots, k + 2s\}, \quad 0 \le k \le N_j - 2s - 2,$

b) $K^+_{j,k} = \{0, \ldots, N_j - 1\}, \quad N_j - 2s - 1 \le k \le 2s,$

c) $K^+_{j,k} = \{k - 2s, \ldots, N_j - 1\}, \quad 2s + 1 \le k \le N_j - 1,$

2.) $4s + 2 \le N_j$:

a) $K^+_{j,k} = \{0, \ldots, k + 2s\}, \quad 0 \le k \le 2s,$

b) $K^+_{j,k} = \{k - 2s, \ldots, k + 2s\}, \quad 2s + 1 \le k \le N_j - 2s - 2,$

c) $K^+_{j,k} = \{k - 2s, \ldots, N_j - 1\}, \quad N_j - 2s - 1 \le k \le N_j - 1.$

From this we conclude

$$\check{l}^0_{j,k} = \begin{cases} 0 & , 0 \le k \le 2s \\ k - 2s & , 2s + 1 \le k \le N_j - 1 \end{cases},$$

$$\check{l}^1_{j,k} = \begin{cases} k + 2s & , 0 \le k \le N_j - 2s - 2 \\ N_j - 1 & , N_j - 2s - 1 \le k \le N_j - 1 \end{cases}.$$

Now, condition (3.25) can be written in the form

$$\pi_j(K^+_{j,k}) = \{\lfloor \check{l}^0_{j,k}/2 \rfloor, \ldots, \lfloor \check{l}^1_{j,k}/2 \rfloor\} \subset \{\lfloor k/2 \rfloor - q, \ldots, \lfloor k/2 \rceil + q\} = \mathcal{N}^q_{j-1,\pi_j(k)}$$

which is satisfied if

$$q \ge \max\{\lfloor k/2 \rfloor - \lfloor \check{l}^0_{j,k}/2 \rfloor, \lfloor \check{l}^1_{j,k}/2 \rfloor + \lfloor k/2 \rfloor\} \ge s.$$

□

Another constraint for the grading parameter q might be imposed, if the adaptive grid has to be locally uniform. Then we deduce from Theorem 4 the following result for the curvilinear case. Once more the result is motivated by the illustration in Fig. 3.10 for the one–dimensional case according to Sect. 2.5.2. Here we assume that the stencil of the modified box wavelets is symmetric, i.e., $\mu = 0$ in (2.39).

Corollary 5. *(Locally uniform grid)*
Let the assumptions of Theorem 4 hold. In the curvilinear case the grid is locally uniform of degree p, if the grading parameter q satisfies

$$q \geq \max\left(s, \lceil(\lceil p/2\rceil + s)/2\rceil\right). \tag{3.47}$$

Proof. In order to prove the assertion we have to verify condition (3.35) in Theorem 4. Again we restrict to the one–dimensional case, see Corollary 4. Then the index sets in (3.35) can be represented as

$$\begin{aligned}
\mathcal{N}^p_{j+1,k'} &= \{\max(0, k'-p), \ldots, \min(N_{j+1}-1, k'+p)\}\,,\\
\pi_{j+1}(\mathcal{N}^p_{j+1,k'}) &= \{\max(0, \pi_{j+1}(k'-p)), \ldots, \min(N_j-1, \pi_{j+1}(k'+p))\}\,,\\
\mathcal{N}^q_{j-1,\pi_j(k)} &= \{\max(0, \pi_j(k)-q), \ldots, \min(N_{j-1}-1, \pi_j(k)+q)\}\,,\\
\mathcal{L}^s_{j,r} &= \{l_{j,r}, \ldots, l_{j,r}+2s\}\,,\\
\pi_j(\mathcal{L}^s_{j,r}) &= \{\pi_j(l_{j,r}), \ldots, \pi_j(l_{j,r})+s\}\,,
\end{aligned}$$

where we employ $\pi_j(r) = \lfloor r/2 \rfloor$. Now condition (3.35) reads

$$\bigcup_{r\in\pi_{j+1}(\mathcal{N}^p_{j+1,k'})} \pi_j(\mathcal{L}^s_{j,r}) \subset \mathcal{N}^q_{j-1,\pi_j(k)}.$$

From this we deduce

$$\max(0, \pi_j(k)-q) \leq \pi_j(l_{j,r}) \leq \min(N_{j-1}-1, \pi_j(k)+q) - s \tag{3.48}$$

for all $r \in \pi_{j+1}(\mathcal{N}^p_{j+1,k'})$. Since the coefficients $l_{j,r}$ are monotonously increasing, i.e., $l_{j,r} \leq l_{j,r'}$ for $r \leq r'$, we only have to check

$$\begin{aligned}
&\pi_j(l_{j,r_-}) \geq \max(0, \pi_j(k)-q) \text{ and}\\
&\pi_j(l_{j,r_+}) \leq \min(N_{j-1}-1, \pi_j(k)+q) - s
\end{aligned}$$

with $r_- := \max(0, \pi_{j+1}(k'-p))$ and $r_+ := \min(N_j-1, \pi_{j+1}(k'+p))$. We now investigate the inequality (3.48) in detail. To this end, we distinguish three cases depending on k for the estimate to the left and the right, respectively, i.e.,

1) $0 \leq k \leq s \pm \lceil p/2\rceil - 1$,
2) $s \pm \lceil p/2\rceil \leq k \leq N_j - 1 - s \pm \lceil p/2\rceil$,
3) $N_j - s \pm \lceil p/2\rceil \leq k \leq N_j - 1$.

Here the "+" and "−" correspond to the lower (L) and upper (U) estimate in (3.48), respectively. Since $k' = 2k + i_k$ for an appropriate $i_k \in \{0, 1\}$, this results in different bounds for k', namely

1) $0 \leq k' \leq 2(s \pm \lceil p/2\rceil) - 1$,

2) $2(s \pm \lceil p/2 \rceil) \leq k' \leq 2(N_j - s \pm \lceil p/2 \rceil - 1,$

3) $2(N_j - s \pm \lceil p/2) \rceil \leq k' \leq N_{j+1} - 1.$

From this we conclude the upper and lower bounds for r_- and r_+

1) $0 \leq r_\mp \leq s - 1,$ 2) $s \leq r_\mp \leq N_j - 1 - s,$ 3) $N_j - s \leq r_\mp \leq N_j - 1.$

Then we obtain for the coefficients

1) $l_{j,r} = 0,$ 2) $l_{j,r} = r_\mp - s = \lceil (k' \mp p)/2 \rceil - s,$ 3) $l_{j,r} = N_j - 2s - 1.$

Finally, we derive the constraints for the grading parameter

L1 , U3) $q \geq \lfloor (\lceil p/2 \rceil + s - 1)/2 \rfloor = \lceil (\lceil p/2 \rceil + s)/2 \rceil - 1,$

L2 , U2) $q \geq \lceil (\lceil p/2 \rceil + s)/2 \rceil,$

L3 , U1) $q \geq s.$

Now, condition (3.47) is obtained choosing the maximal value on the right hand sides. □

According to (3.47) the grading parameter q is always larger than the stencil width s. This implies that the local multiscale transformation and its inverse are always feasible, see Corollary 4. In general, the requirement of a locally uniform grid imposes a stronger constraint on the grading parameter q than the feasibility of the local transformations. For instance, in case of stencil width $s = 1$ and grid grading degree $p = 3$ the grading degree of the significant details has to be larger than $q = 2$.

On the other hand, the grading of the adaptive grid implies the grading of the significant details, see Proposition 2. In the curvilinear case we can specify condition (3.8).

Corollary 6. *(Graded grid ⇒ graded tree)*
Let the adaptive grid be graded of degree p. Then the tree of significant details is graded of degree q provided that

$$0 \leq q \leq \lfloor p/4 \rfloor. \tag{3.49}$$

Proof. Again we consider only the one–dimensional case. Then we have to determine the sets which are involved in (3.8), i.e.

$$\begin{aligned}
\mathcal{N}^q_{j-1,\pi_j(k)} &= \{\max(0, k'' - q), \ldots, \min(N_{j-1} - 1, k'' + q)\} \\
\mathcal{N}^p_{j+1,k'} &= \{\max(0, 2k + i - p), \ldots, \min(N_{j+1} - 1, 2k + i + p)\} \\
\pi_{j+1}(\mathcal{N}^p_{j+1,k'}) &= \{\max(0, k^-_{i,p}, \ldots, \min(N_j - 1, k^+_{i,p}\} \\
\pi_j(\pi_{j+1}(\mathcal{N}^p_{j+1,k'})) &= \{\max(0, k'' + l^-_{i,p}), \ldots, \min(N_{j-1} - 1, k'' + l^+_{i,p})\}
\end{aligned}$$

with

$$k' = 2k + i = 4k'' + 2l + i, \ i, l \in \{0, 1\}$$

and

$$k_{i,p}^{\pm} := k + \lfloor (i \pm p)/2 \rfloor, \quad l_{i,p}^{\pm} := \lfloor (l + \lfloor (i - p)/2 \rfloor)/2 \rfloor.$$

From this we immediately conclude the assertion. □

According to Sect. 2.5.2, the resulting condition (3.49) is illustrated for the one–dimensional case, cf. Fig. 3.8. Here we assume that the stencil of the modified box wavelets is symmetric, i.e., $\mu = 0$ in (2.39).

4 Adaptive Finite Volume Scheme

In the previous chapters we have outlined independently the core ingredients of a FVS and the framework of the multiscale setting. Now we are combining these two settings in order to derive an adaptive FVS.

We start with a given reference FVS determined by the averages and flux balances on the finest resolution level. From this scheme we deduce evolution equations for the details and the averages. In order to reduce the complexity of the reference scheme these equations are only performed for the local averages corresponding to the adaptive grid $\tilde{\mathcal{G}}^{n+1}_{L,\varepsilon}$ and the significant details $\tilde{\mathcal{D}}^{n+1}_{L,\varepsilon}$, respectively. These sets are determined by means of the local multiscale transformation of the local averages corresponding to the old time level.

It turns out that the construction of appropriate numerical fluxes on the locally refined grid and the prediction of the significant details on the new time level are the core ingredients which crucially influence the performance as well as the reliability of the adaptive scheme.

By construction, the complexity of the resulting scheme is proportional to $\#\,\mathcal{D}^n_{L,\varepsilon}$ whereas the complexity of the reference FVS on the finest resolution level is proportional to N_L. Hence the reference FVS is significantly accelerated. In particular, the speed–up rates increase asymptotically with increasing number of refinement levels. This is verified by numerical experiments, see Sect. 7.

In the sequel, we first outline the construction of the adaptive FVS and then present the main algorithms.

4.1 Construction

The starting point for the construction of the adaptive FVS is a nested grid hierarchy according to Definition 3. For the finest resolution level L we then consider an arbitrary (explicit) FVS, see Sect. 1.2, that can be written in the form

$$\mathbf{v}^{n+1}_{L,k} = \mathbf{v}^n_{L,k} - \lambda_{L,k}\, \mathsf{B}^n_{L,k}, \quad \lambda_{L,k} := \tau/|V_{L,k}|, \tag{4.1}$$

where the *flux balance* is defined by

$$\mathsf{B}^n_{L,k} := \sum_l |\Gamma^L_{k,l}|\, \mathsf{F}^{L,n}_{k,l} \tag{4.2}$$

and τ is chosen such that the CFL restriction for the explicit FVS holds for the *finest* grid. In order to handle the numerical fluxes at the boundary $\partial\Omega$ we again introduce a layer of ghost cells for each discretization level such that the hierarchy of extended grids is still nested in the sense of Definition 3. Note, that this does not affect the multiscale transformation but only the flux computation in the evolution step.

In addition to the notation introduced in Sect. 1.2 we have to indicate the level. In the sequel, we only consider numerical fluxes that fit into the following setting.

Assumption 2.

(1) The numerical fluxes $\mathsf{F}^{L,n}_{k,l}$ are determined by

$$\mathsf{F}^{L,n}_{k,l} := \mathsf{F}(\mathbf{v}^n_{L,r}\ ;\ r \in \mathcal{F}^L_{k,l}), \tag{4.3}$$

where the index set $\mathcal{F}^L_{k,l} \subset I_L$ denotes an arbitrary but fixed stencil.

(2) If $\Gamma^L_{k,l} \subset \partial\Omega$, the numerical fluxes at the boundary are computed by

$$\mathsf{F}^{L,n}_{k,l} = \frac{1}{|\Gamma^L_{k,l}|} \int_{\Gamma^L_{k,l}} \mathbf{f}_{\mathbf{n}^L_{k,l}(\mathbf{x})}(\mathbf{v}_{k,l})\, d\mathbf{x}, \tag{4.4}$$

where the state $\mathbf{v}_{k,l}$ is either determined by the prescribed boundary conditions or by the attached flow field.

(3) The numerical fluxes are consistent, i.e.,

$$\mathbf{v}^n_{L,r} \equiv \mathbf{v},\ r \in \mathcal{F}^L_{k,l} \ \Rightarrow\ \mathsf{F}^{L,n}_{k,l} = \frac{1}{|\Gamma^L_{k,l}|} \int_{\Gamma^L_{k,l}} \mathbf{f}_{\mathbf{n}^L_{k,l}(\mathbf{x})}(\mathbf{v})\, d\mathbf{x}. \tag{4.5}$$

(4) The numerical fluxes are conservative, i.e.,

$$\mathsf{F}^{L,n}_{k,l} = -\mathsf{F}^{L,n}_{l,k}. \tag{4.6}$$

Next we introduce the averages $\mathbf{v}^n_{j,k}$ and details $\mathbf{d}^n_{j,k,e}$ on the coarser scales $j = 0, \ldots, L-1$ by the definitions

$$\mathbf{v}^n_{j,k} := \sum_{r \in \check{\mathcal{M}}^0_{j,k}} \check{m}^{j,0}_{r,k}\, \mathbf{v}^n_{j+1,r}, \quad k \in I_j, \tag{4.7}$$

$$\mathbf{d}^n_{j,k,e} := \sum_{r \in \mathcal{M}^e_{j,k}} \tilde{m}^{j,e}_{r,k}\, \mathbf{v}^n_{j+1,r}, \quad k \in I_j,\ e \in E^* \tag{4.8}$$

for an arbitrary time level t_n by means of the box function and the modified box wavelet. For the sake of simplicity, we use the mask coefficients

$\tilde{m}_{r,k}^{j,e}$ instead of distinguishing between the initial coefficients $\check{m}_{r,k}^{j,e}$ of the box wavelets and the modification terms corresponding to the coefficients $l_{r,k}^{j,e}$. Applying the multiscale transformation (3.16) — (3.20) we obtain discrete evolution equations for the averages and the details where we employ (4.7) and (4.8) as well as (4.1)

$$
\begin{aligned}
\mathbf{v}_{j,k}^{n+1} &= \mathbf{v}_{j,k}^{n} - \lambda_{j,k}\,\mathsf{B}_{j,k}^{n}, \quad k \in I_j, \\
\mathbf{d}_{j,k,e}^{n+1} &= \mathbf{d}_{j,k,e}^{n} - \sum_{r \in \mathcal{M}_{j,k}^{e}} \lambda_{j+1,r}\,\tilde{m}_{r,k}^{j,e} \mathsf{B}_{j+1,r}^{n}, \; k \in I_j, \; e \in E^*.
\end{aligned} \tag{4.9}
$$

Here, the flux balances $\mathsf{B}_{j,k}^{n}$ are determined by

$$
\mathsf{B}_{j,k}^{n} := \sum_{r \in \mathcal{M}_{j,k}^{0}} \frac{|V_{j,k}|}{|V_{j+1,r}|}\,\check{m}_{r,k}^{j,0}\,\mathsf{B}_{j+1,r}^{n} = \sum_{r \in \mathcal{M}_{j,k}^{0}} \mathsf{B}_{j+1,r}^{n}, \tag{4.10}
$$

where we have used (2.31) and $\lambda_{j,k} := \tau/|V_{j,k}|$.

So far we have only derived evolution equations for the averages and details corresponding to the *full* grids. The adaptive FVS is now determined by a significantly smaller selection of evolution equations corresponding to the adaptive grid $\mathcal{G}_{L,\varepsilon}^{n+1}$, i.e.,

$$
\mathbf{v}_{j,k}^{n+1} = \mathbf{v}_{j,k}^{n} - \lambda_{j,k}\,\mathsf{B}_{j,k}^{n}, \quad (j,k) \in \mathcal{G}_{L,\varepsilon}^{n+1}. \tag{4.11}
$$

Here, the adaptive grid is determined by the set of significant details $\mathcal{D}_{L,\varepsilon}^{n+1}$ according to the Algorithm 2. We are now facing two fundamental problems, namely,

- how to compute the local flux balances $\mathsf{B}_{j,k}^{n}$ without employing the flux balances corresponding to the cells of the finest grid and
- how to determine $\mathcal{D}_{L,\varepsilon}^{n+1}$ without evolving the averages on the globally finest level, applying the multiscale transformation and finally thresholding the details?

In the following two subsections we will discuss these questions.

4.1.1 Strategies for Local Flux Evaluation

According to the definition of the flux balances (4.10) the computation of $\mathsf{B}_{j,k}^{n}$ requires *all* flux balances $\mathsf{B}_{L,k'}^{n}$ corresponding to the cells $V_{L,k'} \subset V_{j,k}$ of the *finest* level. However, exploiting the conservation property of the fluxes (4.6) the right hand side of (4.10) can be simplified. For this purpose, the boundary part $\Gamma_{k,l}^{j}$ is decomposed into smaller parts $\Gamma_{k',l'}^{j'}$ according to the local index sets

$$\mathcal{S}_{k,l}^{j,j'} := \{(k',l')\ ;\ k',l' \in I_{j'},\ \Gamma_{k',l'}^{j'} \subset \Gamma_{k,l}^{j},\ V_{j',l'} \subset V_{j,l}\}, \quad 0 \le j \le j' \le L.$$

Then the numerical fluxes and also the flux balances (4.10) can be rewritten as

$$\mathsf{B}_{j,k}^{n} = \sum_{l} |\Gamma_{k,l}^{j}|\, \mathsf{F}_{k,l}^{j,n} \tag{4.12}$$

with the local numerical fluxes

$$\mathsf{F}_{k,l}^{j,n} = \sum_{(k',l')\in\mathcal{S}_{k,l}^{j,j'}} \mathsf{F}_{k',l'}^{j',n} = \sum_{(k',l')\in\mathcal{S}_{k,l}^{j,L}} \mathsf{F}_{k',l'}^{L,n} = \sum_{(k',l')\in\mathcal{S}_{k,l}^{j,L}} \mathsf{F}(\mathbf{v}_{L,r}^{n}\ ;\ r \in \mathcal{F}_{k',l'}^{L}). \tag{4.13}$$

Here, the flux balance $\mathsf{B}_{j,k}^{n}$ is computed by the numerical fluxes $\mathsf{F}_{k',l'}^{L,n}$ corresponding to the boundary part $\Gamma_{k',l'}^{L} \subset \Gamma_{k,l}^{j} \subset \partial V_{j,k}$. All numerical fluxes corresponding to $\Gamma_{k',l'}^{L}$ inside the cell $V_{j,k}$ cancel each other due to the conservation property (4.10). Thus, (4.12) is preferable to (4.10) with regard to an efficient computation. Furthermore, we observe that the computation of the numerical fluxes $\mathsf{F}_{k',l'}^{L,n}$ requires the averages $\mathbf{v}_{L,r}^{n}$, $r \in \mathcal{F}_{k',l'}^{L} \subset I_L$, on the *finest* level which have to be provided by means of the local inverse two–scale transformation (3.21) — (3.23), see Fig. 4.1 (left). This, in general, inflates the complexity by a logarithmic factor depending on the spatial dimension d. In particular, for a dyadic refinement we obtain $\#\,\mathcal{S}_{k,l}^{j,L} = (2^{d-1})^{L-j}$. In order to treat multidimensional problems a cheaper approximation is desirable.

If the reference FVS is based on numerical fluxes corresponding to a structured grid, one alternative is to make the grid locally uniform, i.e., the grid is locally refined such that the fluxes can be computed to data of the same refinement level, see Fig. 4.1 (middle). For the computation of the numerical fluxes $\mathsf{F}_{k,l}^{j,n}$, $(j,k) \in \mathcal{G}_{L,\varepsilon}^{n+1}$, we have to distinguish three cases, namely,

(i) the neighboring cell $V_{j,l}$ also belongs to the adaptive grid, i.e., $(j,l) \in \mathcal{G}_{L,\varepsilon}^{n+1}$,
(ii) the coarser cell $V_{j-1,\pi_j(l)}$ belongs to the adaptive grid, i.e., $(j-1,\pi_j(l)) \in \mathcal{G}_{L,\varepsilon}^{n+1}$ or, equivalently, $(j-1,\pi_j(l),e) \notin \mathcal{D}_{L,\varepsilon}^{n+1}$, $e \in E^*$, and
(iii) the neighboring cell has been refined due to significant details, i.e., $(j,l,e) \in \mathcal{D}_{L,\varepsilon}^{n+1}$ for some $e \in E^*$.

Here, the grading of the grid implies that the levels of two neighboring cells differ at most by one. In case (i) and (ii) the numerical flux is computed by

$$\mathsf{F}_{k,l}^{j,n} = \mathsf{F}(\mathbf{v}_{j,r}^{n}\ ;\ r \in \mathcal{F}_{k,l}^{j}) \tag{4.14}$$

and in case (iii) by

$$\mathsf{F}_{k,l}^{j,n} = \sum_{(k',l')\in\mathcal{S}_{k,l}^{j,j+1}} \mathsf{F}_{k',l'}^{j+1,n} = \sum_{(k',l')\in\mathcal{S}_{k,l}^{j,j+1}} \mathsf{F}(\mathbf{v}_{j+1,r}^{n}\ ;\ r \in \mathcal{F}_{k',l'}^{j+1}). \tag{4.15}$$

Proceeding in such a way, we eventually have to compute additional averages $\mathbf{v}_{j,r}^{n}$ or $\mathbf{v}_{j+1,r}^{n}$ using (3.21) — (3.23). In general, the numerical fluxes do not only depend on two values corresponding to the two adjacent cells but on a larger stencil, e.g., high–order accurate FVS employing a higher order reconstruction. Therefore the neighborhood of cells to be made locally uniform is larger. To this end, we have introduced the notion of a locally uniform grid of degree p, see Definition 8. Here, the degree p can be ensured if the grid is graded of degree q where q is chosen sufficiently large, see Theorem 4. This again imposes a constraint on the grading parameter for the set of significant details, see Proposition 1.

Another alternative of making the computation more efficient is the use of an unstructured approach. Here the numerical fluxes are computed by means of the averages provided by the adaptive grid, i.e.,

$$\mathsf{F}_{k,l}^{j,n} = \mathsf{F}(\mathbf{v}_{j,r}^n \; ; \; (j,r) \in \mathcal{F}_{k,l}), \quad \mathcal{F}_{k,l} \subset \mathcal{G}_{L,\varepsilon}^{n+1}. \tag{4.16}$$

In this case no local refinements are necessary, see Fig. 4.1 (right).

Although the two alternatives are less expensive than the exact computation according to (4.12) we have to be aware that an error is introduced. As investigations in [CKMP01] show, this error might become significant in case of a first order FVS. However, if the reference FVS is a higher order scheme employing a high–order reconstruction, then no significant loss in accuracy has been observed.

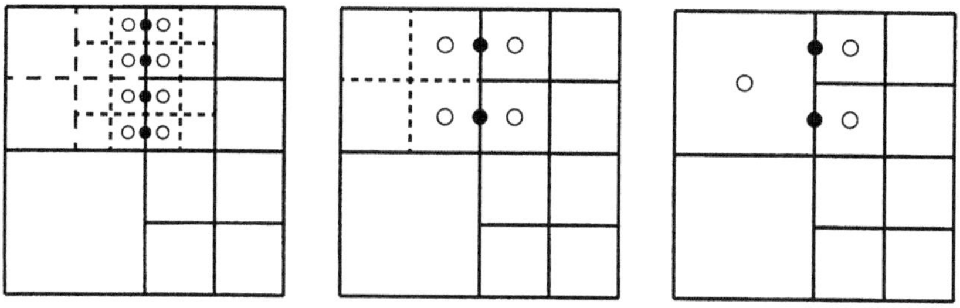

Fig. 4.1. Exact (left), locally structured (middle) and unstructured (right) flux evaluation, (•) numerical fluxes, (∘) cell averages

4.1.2 Strategies for Prediction of Details

The core ingredient of the adaptive FVS is the prediction of the set $\mathcal{D}_{L,\varepsilon}^{n+1}$ by means of the data $\{\mathbf{v}_{j,k}^n\}_{(j,k)\in\mathcal{G}_{L,\varepsilon}^n}$ and $\{\mathbf{d}_{j,k,e}^n\}_{(j,k,e)\in\mathcal{D}_{L,\varepsilon}^n}$, respectively. A first approach was introduced by Harten in [Har94, Har95]. It is based on heuristic arguments employing characteristic features of hyperbolic conservation laws, see also Sect. 1:

- The speed of propagation is finite. This implies that information moves according to the locally finite characteristic speeds, i.e., the eigenvalues of the Jacobian corresponding to the normal fluxes (1.3). The time–space continuum limited by the characteristics propagating with maximal and minimal characteristic speeds form the *range of influence.* For systems of conservation laws this is a cone.
- Discontinuities may develop although the initial data are smooth. In case of *nonlinear* conservation laws this is caused by the intersection of characteristics.

From these observations Harten deduced a prediction set $\tilde{\mathcal{D}}^{n+1}_{L,\varepsilon}$ by means of two criteria applied to each significant detail $(j,k,e) \in \mathcal{D}^n_{L,\varepsilon}$ of the old time level, namely,

(i) details in a local neighborhood of a significant detail may also become significant within one time step,i.e.,

$$\|\mathbf{d}^n_{j,k,e}\|_{\infty,*} \geq \varepsilon_j \quad \Rightarrow \quad (j,r,e) \in \tilde{\mathcal{D}}^{n+1}_{L,\varepsilon},\ r \in \mathcal{N}^{q'}_{j,k},\ e \in E^*, \tag{4.17}$$

and

(ii) gradients may become steeper causing significant details on higher levels, i.e.,

$$\|\mathbf{d}^n_{j,k,e}\|_{\infty,*} \geq 2^{M+1}\,\varepsilon_j \quad \Rightarrow \quad (j+1,r,e) \in \tilde{\mathcal{D}}^{n+1}_{L,\varepsilon},\ r \in \mathcal{M}^0_{j,k},\ e \in E^*. \tag{4.18}$$

In principle, it suffices to choose $q' = 1$, since the CFL restriction for explicit FVS ensures that the time step is sufficiently small such that a perturbation is propagating only from one cell to another. To guarantee a *reliable* adaptive scheme the prediction set $\tilde{\mathcal{D}}^{n+1}_{L,\varepsilon}$ has to satisfy the *reliability condition*

$$\mathcal{D}^n_{L,\varepsilon} \cup \mathcal{D}^{n+1}_{L,\varepsilon} \subset \tilde{\mathcal{D}}^{n+1}_{L,\varepsilon}. \tag{4.19}$$

So far, Harten's approach has not yet been verified to fulfill (4.19). In [CKMP01] a slightly different strategy is proposed. The basic idea is to determine all details $\mathbf{d}^{n+1}_{j',k',e'}$ on the new time level which are influenced by a detail $\mathbf{d}^n_{j,k,e}$ on the old time level. In order to represent this *influence set* $\mathcal{D}_{j,k,e}$ we have to determine all averages $\mathbf{v}^n_{L,r}$, $r \in \Sigma_{j,k,e} \subset I_L$, that are influenced by the detail $\mathbf{d}^n_{j,k,e}$, i.e., the *range of influence* of $\mathbf{d}^n_{j,k,e}$. Accordingly, we determine all averages $\mathbf{v}^{n+1}_{L,r}$, $r \in \tilde{\Sigma}_{j',k',e'} \subset I_L$ on which the detail $\mathbf{d}^{n+1}_{j',k',e'}$ depends, i.e., the *domain of dependence* of $\mathbf{d}^{n+1}_{j',k',e'}$. In order to merge the index sets $\Sigma_{j,k,e}$ and $\tilde{\Sigma}_{j',k',e'}$ which correspond to different time levels, we either have to determine the *backward influence domain* $\tilde{\Sigma}^-_{j',k',e'}$ or the *forward influence domain* $\Sigma^+_{j,k,e}$ by means of the evolution equation on level L. In the sequel we describe how to compute the sets $\Sigma_{j,k,e}$ and $\tilde{\Sigma}_{j',k',e'}$. Since we also need the range of influence and the domain of dependence for

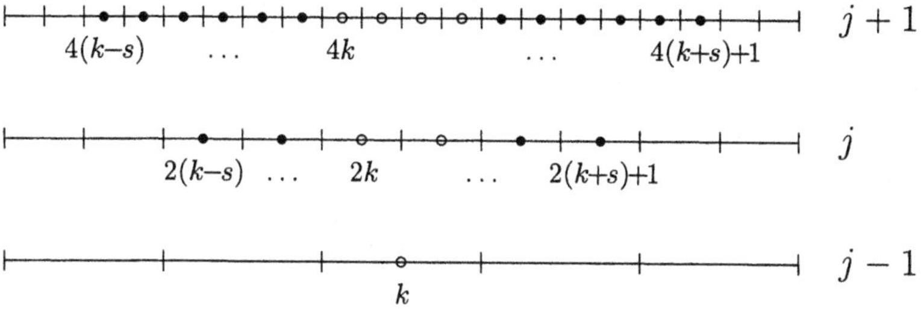

Fig. 4.2. Illustration of the range of influence $\Sigma^{(l)}_{j,k,e}$

the averages $\overline{v}^n_{j,k}$ we introduce the convention $d^n_{j,k,0} := \overline{v}^n_{j,k}$.

Range of influence. The set $\Sigma_{j,k,e}$ is determined by the supports $\mathcal{G}^{*,e}_{j,k}$, $e \in E$, according to the local inverse multiscale transformation (3.21) — (3.23). From these two–scale relations it can be recursively computed

$$\Sigma^{(j)}_{j,k,e} := \{k\},$$

$$\Sigma^{(l+1)}_{j,k,e} := \bigcup\nolimits_{r \in \Sigma^{(l)}_{j,k,e}} \mathcal{G}^{*,e}_{l,r}, \; l = j, \ldots, L-1,$$

$$\Sigma_{j,k,e} := \Sigma^{(L)}_{j,k,e}.$$

Since the supports of the modified box wavelets fulfill $\mathcal{G}^{*,e}_{j,k} = \check{\mathcal{G}}^{*,e}_{j,k} = \check{\mathcal{M}}^0_{j,k}$, $e \in E^*$, and the grids are nested, we may interpret the corresponding range of influence as

$$\Sigma_{j,k,e} = \{r \in I_L \; ; \; V_{L,r} \subset V_{j,k}\}, \quad e \in E^*,$$

i.e., the partition of the cell $V_{j,k}$ by cells on the finest level. This does not hold for $e = 0$. In this case the support corresponding to the index set $\Sigma^{(l)}_{j,k,0}$, i.e., $\bigcup_{r \in \Sigma^{(l)}_{j,k,0}} V_{l,r} \supset V_{j,k}$, is successively growing with increasing level l. This is illustrated in Fig. 4.2 for the one–dimensional case according to Sect. 2.5.2. Here we assume that the stencil of the modified box wavelets is symmetric, i.e., $\mu = 0$ in (2.39). In particular, we obtain $\mathcal{G}^{*,0}_{j,k} = \{2(k-s), \ldots, 2(k+s)+1\}$, cf. Sect. 3.8. The stencils $\Sigma^{(l)}_{j,k,0}$ and $\Sigma^{(l)}_{j,k,1}$ correspond to the cells indicated by both • or ∘ and ∘, respectively.

Domain of dependence. The index set $\tilde{\Sigma}_{j,k,e}$, $e \in E$, can be computed by means of the local multiscale transformation (3.20) and recursively applying the two–scale relation (3.16)

$$\tilde{\Sigma}^{(j)}_{j,k,e} := \mathcal{M}^e_{j,k},$$

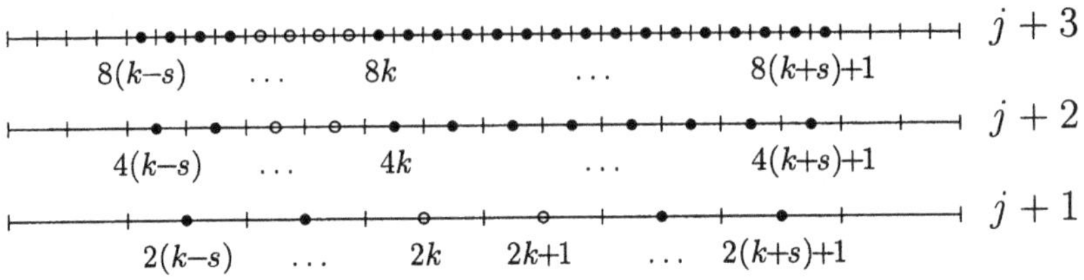

Fig. 4.3. Illustration of the domain of dependence $\tilde{\Sigma}^{(l)}_{j,k,e}$

$$\tilde{\Sigma}^{(l+1)}_{j,k,e} := \bigcup\nolimits_{r\in\tilde{\Sigma}^{(l)}_{j,k,e}} \check{\mathcal{M}}^0_{l+1,r},\ l = j,\dots,L-2,$$

$$\tilde{\Sigma}_{j,k,e} := \Sigma^{(L-1)}_{j,k,e}.$$

Note, that this set corresponds to the support of the modified box wavelet $\psi_{j,k,e}$, $e \in E^*$, according to Algorithm 1, and the box function $\tilde{\varphi}_{j,k} \equiv \psi_{j,k,0}$, respectively, i.e., $\operatorname{supp}\psi_{j,k,e} = \bigcup_{r\in\tilde{\Sigma}_{j,k,e}} V_{L,r}$. Similar to the set $\Sigma_{j,k,e}$, $e \in E^*$, we may interpret the domain of dependence as

$$\tilde{\Sigma}_{j,k,e} = \bigcup_{s\in\mathcal{M}^e_{j,k}} \{r \in I_L \ ;\ V_{L,r} \subset V_{j+1,s}\}, \quad e \in E,$$

i.e., the partition of the cells $V_{j+1,s}$, $s \in \mathcal{M}^e_{j,k}$ by cells on the finest level. In analogy to the range of influence, we consider again the one–dimensional case where we employ the modified box wavelet with symmetric support, cf. Sect. 2.5.2. According to Sect. 3.8 we determine $\mathcal{M}^0_{j,k} = \{2k, 2k+1\}$ and $\mathcal{M}^1_{j,k} = \{2(k-s), 2(k+s)+1\}$. In Fig. 4.3 the index sets $\tilde{\Sigma}^{(l)}_{j,k,e}$ are indicated by $\circ$ $(e=0)$ and $\bullet$ or $\circ$ $(e=1)$, respectively.

Backward (forward) influence domain. By means of the explicit evolution equation (4.1) we know how the averages of two time levels are correlated. For the sake of simplicity, we assume that the discretization stencil which is involved in the computation of the numerical flux balance $\mathsf{B}^n_{L,k}$ and the corresponding numerical fluxes $\mathsf{F}^{L,n}_{k,l}$, see (4.2)and (4.3), is determined by the index set $\mathcal{N}^p_{L,k}$ indicating the neighborhood of degree p corresponding to the cell $V_{L,k}$ on the finest level, i.e., $\bigcup_l \mathcal{F}^L_{k,l} \subset \mathcal{N}^p_{L,k}$. In order to merge the index sets $\Sigma_{j,k,e}$ and $\tilde{\Sigma}_{j',k',e'}$ we have two alternatives based on looking back or forward in time. One possibility is to collect all indices corresponding to averages $\mathbf{v}^n_{L,r}$, $r \in \tilde{\Sigma}^-_{j',k',e'}$, on the old time level which are needed when computing $\mathbf{v}^{n+1}_{L,r}$, $r \in \tilde{\Sigma}_{j',k',e'}$ according to the evolution operator, i.e.,

$$\tilde{\Sigma}^-_{j',k',e'} := \bigcup_{r\in\tilde{\Sigma}_{j',k',e'}} \mathcal{N}^p_{L,r}.$$

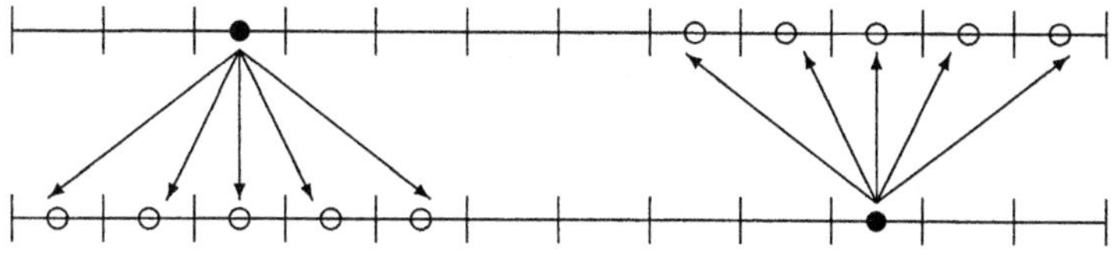

Fig. 4.4. Backward (left) and forward (right) influence domain

Conversely, we can determine the numerical range of influence corresponding to the averages $\mathbf{v}_{l,r}^n$, $r \in \Sigma_{j,k,e}$, i.e.,

$$\Sigma_{j,k,e}^+ := \{r \in I_L \ ; \ \mathcal{N}_{L,r}^p \cap \Sigma_{j,k,e} \neq \emptyset\}.$$

The different points of view are illustrated in Fig. 4.4.

Influence set. The influence set $\mathcal{D}_{j,k,e}$ of details $\mathbf{d}_{j',k',e'}^{n+1}$ which depend on the detail $\mathbf{d}_{j,k,e}^n$ and the coarse–scale averages $\overline{v}_{0,k}^n = d_{0,k,0}^n$ can be represented as

$$\begin{aligned} \mathcal{D}_{j,k,e} &= \{(j',k',e') \in \mathcal{D}_{L,0} \ ; \ \tilde{\Sigma}_{j',k',e'}^- \cap \Sigma_{j,k,e} \neq \emptyset\} \\ &= \{(j',k',e') \in \mathcal{D}_{L,0} \ ; \ \tilde{\Sigma}_{j',k',e'} \cap \Sigma_{j,k,e}^+ \neq \emptyset\}. \end{aligned}$$

To realize an efficient implementation we have to determine a priori those indices (j',k',e'), which yield a contribution to $\mathcal{D}_{j,k,e}$.

We note that the detail $\mathbf{d}_{j,k,e}^n$ may not only cause a perturbation in the neighborhood of the cell $V_{j,k}$ similar to (4.17) but may also influence details $\mathbf{d}_{j',k',e'}^{n+1}$ on higher scales. In contrast to Harten's strategy, here $j' > j+1$ is also admissible. Since the additional higher levels inflate the influence set, we would like to bound the number of higher levels to a minimum number which still provides the reliability property (4.19). For this purpose we fix some $\sigma > 1$ which will depend on the smoothness of the *primal* wavelets, see Assumption 4 in Section 5.3. We now assign to each significant detail or coarse–scale average corresponding to $(j,k,e) \in \mathcal{D}_{L,\varepsilon}^n$ and $(0,k,0) \equiv (0,k)$, respectively, a unique index $\nu = \nu(j,k,e)$ such that

$$2^{\nu(j,k,e)\,\sigma}\,\varepsilon_j < \|\mathbf{d}_{j,k,e}^n\|_{\infty,*} \leq 2^{(\nu(j,k,e)+1)\,\sigma}\,\varepsilon_j. \tag{4.20}$$

Since the index $\nu(j,k,e)$ becomes the smaller the larger σ is, it is convenient to choose σ as large as possible. Then the modified prediction set is determined by

$$\tilde{\mathcal{D}}_{L,\varepsilon}^{n+1} := \mathcal{D}_{L,\varepsilon}^n \cup \bigcup_{(j,k,e)\in\overline{\mathcal{D}}_{L,\varepsilon}^n} \{(j',k',e') \in \mathcal{D}_{j,k,e} \ ; \ j' \leq j+\nu(j,k,e)\} \tag{4.21}$$

with $\overline{\mathcal{D}}^n_{L,\varepsilon} := \mathcal{D}^n_{L,\varepsilon} \cup \{(0,k,0)\ ;\ k \in I_0\}$. We would like to remark that the definition of $\tilde{\mathcal{D}}^{n+1}_{L,\varepsilon}$ is slightly different to that in [CKMP01], since we apply *no* threshold on the coarse–scale averages. This would destroy the conservation property of the adaptive scheme, see Lemma 5.

For a scalar one–dimensional conservation law it is possible to estimate the details $\mathbf{d}^{n+1}_{j',k',e'}$ on the new time level by the details $\mathbf{d}^n_{j,k,e}$ on the old time level, see Sect. 5.3, from which we conclude the reliability property (4.19).

Finally, we remark that we only apply Harten's strategy for our computations presented in Sect. 7, because it is more efficient and easier to implement. This decision is motivated by the investigations in [CKMP01].

4.2 Algorithms: Initial data, Prediction, Fluxes and Evolution

In the sequel, we present the algorithmic ingredients of the adaptive FVS. Four basic routines are needed, namely, (i) the multiscale decomposition of the initial data, (ii) the prediction sets $\tilde{\mathcal{D}}^{n+1}_{L,\varepsilon}$, (iii) the computation of the local numerical fluxes and (iv) the evolution of the averages corresponding to the adaptive grid $\mathcal{G}^{n+1}_{L,\varepsilon}$.

First of all, we consider the computation of the set $\mathcal{D}^0_{L,\varepsilon}$. In principle, we have to perform the *full* multiscale decomposition according to (2.40). However, this would require the computation of *all* averages on the finest grid. In order to reduce the computational complexity this has to be avoided. Therefore we present an algorithm that proceeds levelwise from coarse to fine scale according to Algorithm 2. The basic idea is as follows: first we compute all averages on level 1, i.e., the adaptive grid is assumed to be the grid of level 1, and determine the details by the local two–scale transformation (3.17) — (3.20). If there is at least one significant detail corresponding to a cell $V_{0,k}$ then this cell has actually to be refined otherwise the adaptive grid is locally coarsened. After having handled all existing cells of this level we then proceed to the next higher level where we refine again the cells of the locally highest level and determine the details by the local two–scale transformation. These details determine which cells actually remain. This process is successively repeated until the a priorly fixed highest refinement level is reached. As input data the algorithm needs (i) the number of refinement levels L, (ii) the sequence of threshold values ε, (iii) the scaling parameter *tol* and (iv) the initial function u_0. Then the algorithm provides the set $\mathcal{D}^0_{L,\varepsilon}$ of significant details corresponding to the initial data u_0. Note, that this set is not necessarily graded. Moreover, it is not guaranteed from an analytical point of view that this set actually contains *all* significant details. However, in many applications the initial data are given by piecewise constant data. In this case the set is supposed to be reliable.

Algorithm 7. *(Multiscale decomposition of initial data)*
$I_0^+ := I_0$
<u>for</u> $j = 0$ <u>to</u> *L-1* <u>do</u>

1. $U_j := I_j^+ \cup \bigcup_{e \in E^*} \bigcup_{k \in I_j^+} \mathcal{L}_{j,k}^e$

2. $R_{j+1} := \bigcup_{k \in U_j} \check{\mathcal{M}}_{j,k}^0$

3. $\hat{u}_{j+1,k} := \frac{1}{|V_{j+1,k}|} \int_{V_{j+1,k}} u_0(\mathbf{x})\, d\mathbf{x},\ k \in R_{j+1}$

4. $\hat{u}_{j,k} = \sum_{r \in \check{\mathcal{M}}_{j,k}^0} \check{m}_{r,k}^{j,0}\, \hat{u}_{j+1,r},\ k \in U_j \backslash I_j^+$

5. $d_{j,k,e} = \sum_{r \in \mathcal{L}_{j,k}^e} l_{r,k}^{j,e}\, \hat{u}_{j,r} + \sum_{r \in \check{\mathcal{M}}_{j,k}^e} \check{m}_{r,k}^{j,e}\, \hat{u}_{j+1,r}, \quad e \in E^*,\ k \in I_j^+$

6. $u_{\infty,i}^{j+1} := \max_{k \in R_{j+1}} |(\hat{u}_{j+1,k})_i|,\ i = 1, \ldots, m$

7. <u>for</u> $e \in E^*,\ k \in I_j^+$ <u>do</u>
7.1 <u>for</u> $i = 1$ <u>to</u> m <u>do</u>

<u>if</u> $u_{\infty,i}^{j+1} < tol$

<u>then</u> $(d_{j,k,e})_i^* := (d_{j,k,e})_i$ <u>else</u> $(d_{j,k,e})_i^* := (d_{j,k,e})_i / u_{\infty,i}^{j+1}$

7.2 $\|d_{j,k,e}\|_{\infty,*} := \|d_{j,k,e}^*\|_\infty$

7.3 <u>if</u> $\|d_{j,k,e}\|_{\infty,*} > \varepsilon_j$ <u>then</u> $(k,e) \in J_{j,\varepsilon}$ <u>else</u> *delete* $d_{j,k,e}$

8. $I_{j+1}^+ := \bigcup_{(k,e) \in J_{j,\varepsilon}} \mathcal{M}_{j,k}^e$

Firstly, we have to initialize the set I_0^+ of cells on level 0 for which we have to compute the details. In order to determine the significant details we proceed levelwise starting on the coarsest level. For each level we have to perform the following steps. First, we determine all indices of cells on the locally highest level j which have to be refined, see step 1, in view of the local two–scale transformation. To this end, the set I_j^+ has to be inflated, since additional averages $\hat{u}_{j,k}$ have to be computed in order to perform (3.17) in step 5. Then the indices of the new cells on the next finer level $j+1$ and the corresponding averages are computed, see step 2 and 3. Before performing the local two–scale decomposition (3.17) — (3.20) in step 5, we first have to provide the averages $\hat{u}_{j,k}$ not yet determined from the data on the finer level, see step 4. Similar to Algorithm 4 we perform the thresholding and determine $\mathcal{D}_{L,\varepsilon}^0$, see step 6 and 7. By means of the significant details we deduce the cells of level $j+1$ for which we have to determine the details in the next iteration, see step 8. Finally we remark that all averages $\hat{u}_{0,k}$, $k \in I_0$, have been determined when performing the loop for $j = 0$, since $I_0^+ = I_0$ and, hence, $U_0 = I_0$. These data also have to be provided by the multiscale decomposition of the initial data.

In order to predict the adaptive grid $\mathcal{G}_{L,\varepsilon}^{n+1}$ on the next time level we have to determine the set $\tilde{\mathcal{D}}_{L,\varepsilon}^{n+1}$. According to Sect. 4.1.2 two alternatives are at hand. Here, we only outline Harten's strategy which has been applied for our computations in Sect. 7. As input data we need (i) the number of refinement levels L, (ii) the sets $\mathcal{D}_{L,\varepsilon}^n$ and $\mathcal{J}_{j,\varepsilon}^n$, $j = 0, \ldots, L-1$, respectively, of significant details and (iii) the sequences of significant details $d_{j,k,e}^n$, $(j,k,e) \in \mathcal{D}_{L,\varepsilon}^n$ corresponding to the old time level t_n. Then the algorithm provides $\tilde{\mathcal{D}}_{L,\varepsilon}^{n+1}$ and $\tilde{\mathcal{J}}_{j,\varepsilon}^{n+1}$, $j = 0, \ldots, L-1$, according to (4.17) and (4.21), respectively.

Algorithm 8. *(Prediction due to Harten)*
$\tilde{\mathcal{J}}_{j,\varepsilon}^{n+1} := \emptyset, \quad j = 0, \ldots, L-1$
<u>for</u> $j = 0$ <u>to</u> L-1 <u>do</u>

1. <u>for</u> $(j,k,e) \in \mathcal{J}_{j,\varepsilon}^n$ <u>do</u>
1.1 $\tilde{\mathcal{J}}_{j,\varepsilon}^{n+1} := \tilde{\mathcal{J}}_{j,\varepsilon}^{n+1} \cup \{(j,r,e) \ ; \ r \in \mathcal{N}_{j,k}^{q'}, \ e \in E^*\}$

1.2 <u>if</u> $\|\mathbf{d}_{j,k,e}^n\|_{\infty,*} \geq 2\varepsilon_j$ <u>then</u>

$$\tilde{\mathcal{J}}_{j+1,\varepsilon}^{n+1} := \tilde{\mathcal{J}}_{j+1,\varepsilon}^{n+1} \cup \{(j+1,r,e) \ ; \ r \in \mathcal{M}_{j,k}^0, \ e \in E^*\}$$

2. $d_{j,k,e} := 0, \quad (j,k,e) \in \tilde{\mathcal{J}}_{j,\varepsilon}^{n+1} \backslash \mathcal{J}_{j,\varepsilon}^n$

3. $d_{j+1,k,e} := 0, \quad (j+1,k,e) \in \tilde{\mathcal{J}}_{j+1,\varepsilon}^{n+1} \backslash \mathcal{J}_{j+1,\varepsilon}^n$

Note, that we ensure $\mathcal{J}_{j,\varepsilon}^n \subset \tilde{\mathcal{J}}_{j,\varepsilon}^{n+1}$ in step 1.

Next we describe the algorithms for the computation of the numerical fluxes. Here we have to distinguish between the internal and the boundary fluxes, see (4.3) and (4.4). In both cases we need the following input data: (i) the number of refinement levels L, (ii) the set $\tilde{\mathcal{G}}_{L,\varepsilon}^{n+1}$ and $\tilde{\mathcal{I}}_{j,\varepsilon}^{n+1}$, $j = 0, \ldots, L$, respectively, (iii) the corresponding local averages $\mathbf{v}_{j,k}^n$, $(j,k) \in \tilde{\mathcal{G}}_{L,\varepsilon}^{n+1}$, corresponding to the *old* time level t_n and (iv) for the boundary fluxes we have to provide in addition the prescribed boundary data. From this information the numerical fluxes corresponding to the flux balances $\mathsf{B}_{j,k}^n$, $(j,k) \in \tilde{\mathcal{G}}_{L,\varepsilon}^{n+1}$ are determined.

First of all, we consider the computation of the boundary fluxes. To this end we introduce the index set I_j°, $j = 0, \ldots, L$, of the ghost cells in order to handle the edges $\Gamma_{k,l}^j \subset \partial\Omega$, $l \in I_j^\circ$.

Algorithm 9. *(Boundary fluxes)*
<u>for</u> $(j,k) \in \tilde{\mathcal{G}}_{L,\varepsilon}^{n+1}$, $\partial V_{j,k} \cap \partial\Omega \neq \emptyset$ <u>do</u>

<u>for</u> $l \in I_j^\circ$, $\Gamma_{k,l}^j \subset \partial\Omega$ <u>do</u>

1. <u>for</u> $(k',l') \in \mathcal{S}_{k,l}^{j,L}$ <u>do</u>

1.1 determine the state $\mathbf{v}_{k',l'}$ *from the boundary values or the attached flow field*

1.2 $\mathsf{F}^{L,n}_{k',l'} = \frac{1}{|\Gamma^L_{k',l'}|} \int_{\Gamma^L_{k',l'}} \mathbf{f}_{\mathbf{n}^L_{k',l'}(\mathbf{x})}(\mathbf{v}_{k',l'})\, d\mathbf{x}$

2. $\mathsf{F}^{j,n}_{k,l} := \sum_{(k',l')\in\mathcal{S}^{j,L}_{k,l}} \mathsf{F}^{L,n}_{k',l'}$

For the boundary fluxes appropriate states $\mathbf{v}_{k',l'}$ have to be determined from the boundary conditions or the attached flow field. In the latter case, we eventually need the averages on the finest level. Then we have to apply the local inverse two–scale transformation (3.21) — (3.23) recursively. According to (4.4) the boundary fluxes correspond to the *finest* level in order to avoid inconsistency. This is a serious demand, in particular, if the boundary $\partial\Omega$ is curved.

According to Sect. 4.1.1 there are three alternatives for the computation of the internal fluxes.

Algorithm 10. *(Exact flux evaluation)*
<u>for</u> $j = L$ <u>downto</u> *0* <u>do</u>

<u>for</u> $k \in \tilde{\mathcal{I}}^{n+1}_{j,\varepsilon}$ <u>do</u>

<u>for</u> $l \in I_j$, $\Gamma^j_{k,l} \neq \emptyset$, $\Gamma^j_{k,l} \cap \partial\Omega = \emptyset$ <u>do</u>

1. <u>for</u> $(k',l') \in \mathcal{S}^{j,L}_{k,l}$ <u>do</u>

1.1 reconstruct the states $\mathbf{v}^n_{L,r}$, $r \in \mathcal{F}^L_{k',l'}$, *by means of recursively applying the local inverse two–scale transformation (3.21) — (3.23), if not yet available*

1.2 $\mathsf{F}^{L,n}_{k',l'} = \mathsf{F}(\mathbf{v}^n_{L,r}\ ;\ r \in \mathcal{F}^L_{k',l'})$

2. $\mathsf{F}^{j,n}_{k,l} := \sum_{(k',l')\in\mathcal{S}^{j,L}_{k,l}} \mathsf{F}^{L,n}_{k',l'}$

Here, we proceed from fine to coarse. For each cell $V_{j,k}$ of the adaptive grid we determine the numerical fluxes at the interfaces $\Gamma^j_{k,l}$ to the neighboring cells $V_{j,l}$. According to (4.12) the numerical flux $\mathsf{F}^{j,n}_{k,l}$ is the sum of numerical fluxes corresponding to the partition $\Gamma^j_{k,l} = \bigcup_{(k',l')\in\mathcal{S}^{j,L}_{k,l}} \Gamma^L_{k',l'}$, see (4.13). The computation of the internal fluxes $\mathsf{F}^{L,n}_{k',l'}$, i.e., $\Gamma^L_{k',l'} \subset \Omega$, depends on the averages $\mathbf{v}^n_{L,r}$, $r \in \mathcal{F}^L_{k',l'} \subset I_L$, on the finest level. If these data are not available, i.e., $(L,r) \notin \tilde{\mathcal{I}}^{n+1}_{L,\varepsilon}$, these values have to be reconstructed locally. For the boundary fluxes appropriate states $\mathbf{v}_{k',l'}$ have to be determined from the boundary conditions or the attached flow field.

In order to avoid the local reconstruction of values on the full *finest* level, the second strategy is based on the idea to make the grid locally uniform. Then we can still use a numerical flux based on a structured grid, see (4.14) and (4.15). In addition to the above input data we need that the adaptive grid can be made locally uniform of degree p where p is determined by the minimal degree q of neighborhoods $\mathcal{N}^q_{j,k}$ such that $\bigcup_l \mathcal{F}^j_{k,l} \subset \mathcal{N}^q_{j,k}$.

Algorithm 11. *(Locally structured flux evaluation)*
for $j = L$ downto *0* do

for $k \in \tilde{\mathcal{I}}^{n+1}_{j,\varepsilon}$ do
for $l \in I_j$, $\Gamma^j_{k,l} \neq \emptyset$, $\Gamma^j_{k,l} \cap \partial\Omega = \emptyset$ do

if (j,l), $(j-1, \pi_j(l)) \notin \tilde{\mathcal{G}}^{n+1}_{L,\varepsilon}$ then
1. for $(k', l') \in \mathcal{S}^{j,j+1}_{k,l}$ do
1.1 reconstruct the states $\mathbf{v}^n_{j+1,r}$, $r \in \mathcal{F}^{j+1}_{k',l'} \subset I_{j+1}$, *by means of the local inverse two-scale transformation (3.16) — (3.23), if not yet available*

1.2 $\mathsf{F}^{j+1,n}_{k',l'} = \mathsf{F}(\mathbf{v}^n_{j+1,r}\ ;\ r \in \mathcal{F}^{j+1}_{k',l'})$

2. $\mathsf{F}^{j,n}_{k,l} := \sum_{(k',l') \in \mathcal{S}^{j,j+1}_{k,l}} \mathsf{F}^{j+1,n}_{k',l'}$
else
1. reconstruct the states $\mathbf{v}^n_{j,r}$, $r \in \mathcal{F}^j_{k,l} \subset I_j$, *by means of the local inverse two-scale transformation (3.21) — (3.23) or local two-scale transformation (3.16), if not yet available*

2. $\mathsf{F}^{j,n}_{k,l} = \mathsf{F}(\mathbf{v}^n_{j,r}\ ;\ r \in \mathcal{F}^j_{k,l})$

In comparison to Algorithm 10 we proceed differently in case of the internal fluxes. Here we locally refine the neighboring cells at most once. This reduces the reconstruction costs which become significant for higher dimensional problems.

The third strategy is based on an unstructured flux evaluation, see (4.16), where we only access data of the adaptive grid. In principle, this grid need not be graded in view of the flux evaluation.

Algorithm 12. *(Unstructured flux evaluation)*
for $j = L$ downto *0* do

for $k \in \tilde{\mathcal{I}}^{n+1}_{j,\varepsilon}$ do
for $l \in I_j$, $\Gamma^j_{k,l} \neq \emptyset$, $\Gamma^j_{k,l} \cap \partial\Omega = \emptyset$ do
1. determine the stencil $\mathcal{F}_{k,l} \subset \tilde{\mathcal{G}}^{n+1}_{L,\varepsilon}$

2. $\mathsf{F}^{j,n}_{k,l} = \mathsf{F}(\mathbf{v}^n_{j',k'}\ ;\ (j',k') \in \mathcal{F}_{k,l})$

Now the algorithm for the evolution step according to (4.11) can be described. Here we need (i) the number of refinement levels L, (ii) the set $\tilde{\mathcal{G}}^{n+1}_{L,\varepsilon}$ and $\tilde{\mathcal{I}}^{n+1}_{j,\varepsilon}$, $j = 0, \ldots, L$, respectively, corresponding to the adaptive grid, (iii) the corresponding local averages $\mathbf{v}^n_{j,k}$, $(j,k) \in \tilde{\mathcal{G}}^{n+1}_{L,\varepsilon}$ and (iv) the numerical fluxes corresponding to the flux balances $\mathsf{B}^n_{j,k}$, $(j,k) \in \tilde{\mathcal{G}}^{n+1}_{L,\varepsilon}$. Then the local averages $\mathbf{v}_{j,k}^{n+1}$, $(j,k) \in \tilde{\mathcal{G}}_{L,\varepsilon}^{n+1}$ on the new time level t_{n+1} are determined as follows.

Algorithm 13. *(Evolution)*
<u>for</u> $j = L$ <u>downto</u> *0* <u>do</u>

<u>for</u> $k \in \tilde{\mathcal{I}}_{L,\varepsilon}^{n+1}$ <u>do</u>
1. $\mathsf{B}_{j,k}^{n} := \sum_l |\Gamma_{k,l}^{j}| \, \mathsf{F}_{k,l}^{j,n}$

2. $\mathbf{v}_{j,k}^{n+1} = \mathbf{v}_{j,k}^{n} - \lambda_{j,k} \, \mathsf{B}_{j,k}^{n}$

Finally, we summarize the algorithm for the adaptive FVS. The input data are (i) the number of refinement levels L, (ii) the sequence of threshold values ε, (iii) the scaling parameter *tol*, (iv) the grading parameter q, (v) the temporal step size τ and (vi) the initial function u_0. Then the algorithm provides the local averages $\mathbf{v}_{j,k}^{N_T}$, $(j,k) \in \tilde{\mathcal{G}}_{L,\varepsilon}^{N_T}$, corresponding to the time $T = \tau \, N_T$.

Algorithm 14. *(Adaptive finite volume scheme)*
<u>for</u> $n = 0$ <u>to</u> $N_T - 1$ <u>do</u>

1. perform the local multiscale transformation according to Algorithm 3 ($n > 0$) or determine the multiscale decomposition of the initial function u_0 by Algorithm 7 ($n = 0$);

2. threshold the sequence of details by Algorithm 4, i.e., determine the set of significant details $\mathcal{D}_{L,\varepsilon}^{n}$;

3. predict the set of significant details $\tilde{\mathcal{D}}_{L,\varepsilon}^{n+1}$ on the next time level by Algorithm 8 respectively;

4. grade the set of significant details $\tilde{\mathcal{D}}_{L,\varepsilon}^{n+1}$ by Algorithm 5;

5. perform the local inverse multiscale transformation by Algorithm 6;

6. determine the adaptive grid $\tilde{\mathcal{G}}_{L,\varepsilon}^{n+1}$ by Algorithm 2;

7. compute the internal numerical fluxes by Algorithms 10, 11 and 12, respectively, and the boundary fluxes by Algorithm 9;

8. perform the evolution step by Algorithm 4.

Note, that the local (inverse) multiscale transformation is only feasible if the tree of details is sufficiently graded. Moreover, the flux evaluation according to Algorithm 11 requires that the adaptive grid can be made locally uniform. To this end, an additional constraint on the grading parameter q has to be taken into account. And finally, we remark that the CFL restriction for explicit FVS has to hold for the *finest* grid. This is still a bottleneck in the performance of the adaptive scheme, see Sect. 7 for details.

5 Error Analysis

The objective of the proposed adaptive scheme is to reduce for a given FVS computational cost and memory requirements while preserving the accuracy of the reference scheme. In order to quantify this we introduce the averages $\hat{\mathbf{u}}_L^n$ of the exact solution, the averages $\mathbf{v}_L^n$ determined by the FVS and the averages $\overline{\mathbf{v}}_L^n$ of the adaptive scheme. Note, that all sequences correspond to the finest discretization level indicated by L. In particular, the sequence $\overline{\mathbf{v}}_L^n$ is *not* generated by the adaptive scheme on the full uniform highest level L but these data are determined by the adaptive array $\{\mathbf{v}_{j,k}^n\}_{(j,k)\in\mathcal{G}_{L,\varepsilon}^n}$ and could be retrieved from them by means of the local inverse multiscale transformation (3.21) — (3.23) where the non–significant details $d_{j,k,e}$ in (3.23) are put to zero.

An ideal strategy would be to prescribe an error tolerance *tol*. Then the number of refinement levels L should be determined during the computation such that the error meets the tolerance, i.e.,

$$||\hat{\mathbf{u}}_L^n - \overline{\mathbf{v}}_L^n|| \le tol$$

for possibly small L. Here $||\cdot||$ denotes an appropriate norm to be specified below. For this purpose we need an error estimator which is not available for the adaptive scheme at hand. In the sequel, we therefore proceed differently. To this end we split the error into two parts corresponding to the *discretization error* $\tau_L^n := \hat{\mathbf{u}}_L^n - \mathbf{v}_L^n$ of the reference FVS and the *perturbation error* $\mathbf{e}_L^n := \mathbf{v}_L^n - \overline{\mathbf{v}}_L^n$, i.e.,

$$||\hat{\mathbf{u}}_L^n - \overline{\mathbf{v}}_L^n|| \le ||\tau_L^n|| + ||\mathbf{e}_L^n|| \le tol. \tag{5.1}$$

We now assume that there is an a priori error estimate of the discretization error, i.e., $\tau_L^n \sim h_L^\alpha$ where $h_L := \max_{k\in I_L} \operatorname{diam}(|V_{L,k}|)$ denotes the spatial step size and α the order of convergence. Then we ideally would determine the number of refinement levels L such that $h_L^\alpha \sim tol$. In order to preserve the accuracy of the reference FVS we now may admit a perturbation error which is proportional to the discretization error, i.e., $||\mathbf{e}_L^n|| \sim ||\tau_L^n||$. From this we conclude

$$L = L(tol, \alpha) \quad \text{and} \quad \varepsilon = \varepsilon(L). \tag{5.2}$$

For scalar conservation laws error estimates have been derived, see [CCL94, Noe96]. Therefore it remains to verify that the perturbation error can be controlled. Note, that in each time step we introduce an error due to the threshold procedure. Obviously, this error accumulates in each step, i.e., the best we can hope for is an estimate of the form

$$\|\mathbf{e}_L^n\| \leq C\, n\, \varepsilon.$$

However, the threshold error may be amplified in addition by the evolution step. In order to control the cumulative perturbation error we have to prove that the constant C is independent of L, n, τ and ε. For this purpose we will consider the following issues in more detail, namely,

i. the uniform boundedness of the perturbation error,
ii. the stability of the threshold procedure and
iii. the reliability of the prediction procedure.

5.1 Perturbation Error

In the sequel, we investigate the difference between the averages $\mathbf{v}_L^n$ of the reference FVS and the averages $\overline{\mathbf{v}}_L^n$ of the adaptive scheme. The error is measured in the weighted l^1–metric

$$\|\mathbf{u}_L\|_{1,L} := \sum_{k \in I_L} |V_{L,k}|\, |u_{L,k}|,$$

which is equivalent to the L^1–norm of a piecewise constant function. This metric is usually applied in the error analysis of FVS. According to Harten's analysis in [Har95] we will split the error into its different contributions corresponding to thresholding and prediction as well as the local flux approximation. To this end, we introduce several operators that simplify the representation of the error analysis:

- The threshold operator $\mathcal{T}_{\mathcal{D}} : \mathsf{R}^{N_L} \to \mathsf{R}^{N_L}$ is determined by means of the multiscale decomposition $\mathbf{d}^{(L)}$. It puts all non–significant details to zero, i.e., $d_{j,k,e} := 0$, $(j,k,e) \notin \mathcal{D}$, and retains all significant details. Here $\mathcal{D} \subset J^{(L)}$ denotes an arbitrary subset of all details.
- The nonlinear approximation operator $\mathcal{A}_{\mathcal{D}} : \mathsf{R}^{N_L} \to \mathsf{R}^{N_L}$ is defined by $\mathcal{A}_{\mathcal{D}} := \mathcal{M}_L^{-1} \mathcal{T}_{\mathcal{D}} \mathcal{M}_L$. First the sequence of averages is decomposed into its multiscale decomposition. Then the thresholding is performed by means of the set $\mathcal{D}$ of significant details and, finally, the modified averages corresponding to the sequence of thresholded details are determined where we apply the inverse multiscale transformation.
- The evolution operator $\mathcal{E}_L : \mathsf{R}^{N_L} \to \mathsf{R}^{N_L}$ of the reference FVS is determined by (4.1). It evolves the averages from one time level to the next one, i.e., $\mathbf{v}_L^{n+1} = \mathcal{E}_L\, \mathbf{v}_L^n$.

– The evolution operator $\mathcal{E}_{\mathcal{G}} : \mathsf{R}^{N_L} \to \mathsf{R}^{N_L}$ of the adaptive FVS corresponds to the adaptive grid $\mathcal{G} = \mathcal{G}(\mathcal{D})$ and the set of significant details $\mathcal{D}$, respectively. This operator is determined by the evolution equations (4.11). To these equations we apply the local inverse multiscale transformation (3.21)—(3.23) where we put $d_{j,k,e}$ to zero in (3.23), i.e., the local averages on higher scales are reconstructed only from averages on coarser scales corresponding to the adaptive grid. Equivalently, the evolution operator can be derived by putting $d_{j,k,e}^{n+1} := 0$, $(j,k,e) \notin \mathcal{D}$, in (4.9) and then applying the inverse multiscale transformation $\mathcal{M}_L^{-1}$ to the sequence of evolution equations for the multiscale decomposition.

By means of these operators the adaptive FVS mapped onto the highest resolution level can be represented by

$$\overline{\mathbf{v}}_L^{n+1} = \mathcal{M}_L^{-1}\, \mathcal{T}_{\mathcal{D}_{L,\varepsilon}^{n+1}}\, \mathcal{M}_L\, \mathcal{E}_{\tilde{\mathcal{G}}_{L,\varepsilon}^{n+1}} \overline{\mathbf{v}}_L^n = \mathcal{A}_{\tilde{\mathcal{D}}_{L,\varepsilon}^{n+1}}\, \mathcal{E}_{\tilde{\mathcal{G}}_{L,\varepsilon}^{n+1}} \overline{\mathbf{v}}_L^n .$$

Here we assume that the set $\mathcal{D}_{L,\varepsilon}^{n+1}$ is always graded. Otherwise we have to apply an additional grading operator $\mathcal{A}_\varepsilon$ as is done in [CKMP01]. Since the grading does not affect the approximation error, we incorporate this step into the construction of the prediction sets. Note, that for $\varepsilon = 0$ no threshold error is introduced, i.e., $\mathbf{e}_L^n = \mathbf{0}$. Therefore the adaptive scheme may be interpreted as a perturbation of the reference FVS.

Now the following theorem provides an estimate for the perturbation error.

Theorem 5. *(Uniform boundedness of perturbation error)*
Let the following assumptions hold:

(A1) *the approximation error is uniformly bounded, i.e.,*

$$\|\mathbf{u}_L - \mathcal{A}_{\mathcal{D}_{L,\varepsilon}}\, \mathbf{u}_L\|_{1,L} \le C_1\, \varepsilon;$$

(A2) *the local flux approximation is accuracy preserving, i.e.,*

$$\|\mathcal{E}_L \mathbf{v}_L - \mathcal{E}_{\tilde{\mathcal{G}}_{L,\varepsilon}} \mathbf{v}_L\|_{1,L} \le C_2\, \varepsilon;$$

(A3) *the reference FVS is stable, i.e.,*

$$\|\mathcal{E}_L \mathbf{u}_L - \mathcal{E}_L \mathbf{v}_L\|_{1,L} \le (1 + C_3\, \tau)\, \|\mathbf{u}_L - \mathbf{v}_L\|_{1,L};$$

(A4) *the prediction is reliable in the sense of (4.19);*

(A5) *the initial data are consistent, i.e.,* $\|\mathbf{v}_L^0 - \overline{\mathbf{v}}_L^0\|_{1,L} \le C_4\, \varepsilon$.

Then the perturbation error is bounded by

$$\|\mathbf{e}_L^n\|_{1,L} \le C\, \frac{\varepsilon}{\tau} \tag{5.3}$$

for $n\,\tau \le T$ *where* C *is independent of* L, n, τ *and* ε.

Note, that the temporal discretization τ is assumed to be uniform. This is not a severe constraint but simplifies the representation of the analysis.

Proof. In a first step we split the perturbation error into its different contributions corresponding to the stability of the reference FVS (A3), the approximation of the reference numerical fluxes (A2) and the threshold error (A1), i.e.,

$$\|\mathbf{e}_L^n\|_{1,L} \leq \|\mathcal{E}_L \mathbf{v}_L^{n-1} - \mathcal{E}_L \overline{\mathbf{v}}_L^{n-1}\|_{1,L} + a_{n-1} + b_{n-1}$$

with

$$a_{n-1} := \|\mathcal{E}_L \overline{\mathbf{v}}_L^{n-1} - \mathcal{E}_{\tilde{\mathcal{G}}_{L,\varepsilon}^n} \overline{\mathbf{v}}_L^{n-1}\|_{1,L} \leq C_2\, \varepsilon$$

$$b_{n-1} := \|\mathcal{E}_{\tilde{\mathcal{G}}_{L,\varepsilon}^n} \overline{\mathbf{v}}_L^{n-1} - \mathcal{A}_{\mathcal{D}_{L,\varepsilon}^n} \mathcal{E}_{\tilde{\mathcal{G}}_{L,\varepsilon}^n} \overline{\mathbf{v}}_L^{n-1}\|_{1,L} \leq C_1\, \varepsilon.$$

Then we conclude from the assumptions

$$\|\mathbf{e}_L^n\|_{1,L} \leq \|\mathbf{e}_L^{n-1}\|_{1,L}\, (1 + C_3\, \tau) + (C_1 + C_2)\, \varepsilon.$$

By recursion we obtain further

$$\|\mathbf{e}_L^n\|_{1,L} \leq \|\mathbf{e}_L^0\|_{1,L}\, (1 + C_3\, \tau)^n + \varepsilon\, (C_1 + C_2) \sum_{i=0}^{n-1} (1 + C_3\, \tau)^i.$$

We then conclude with $\overline{C} := \max(C_1 + C_2, C_4)$

$$\|\mathbf{e}_L^n\|_{1,L} \leq \varepsilon\, \overline{C}\, \frac{(1 + C_3\, \tau)^{n+1} - 1}{C_3\, \tau} \leq \varepsilon\, \overline{C}\, \frac{e^{C_3\, (n+1)\, \tau} - 1}{C_3\, \tau}$$

in case of $C_3 \neq 0$ and

$$\|\mathbf{e}_L^n\|_{1,L} \leq \varepsilon\, \overline{C}\, \frac{(n+1)\, \tau}{\tau}$$

if the FVS is l^1–contractive, i.e., $C_3 = 0$. Since the maximal number of time steps is bounded by $n \leq T/\tau$ for a bounded time interval $[0, T]$, $T < \infty$, the assertion follows. □

From this theorem we immediately conclude that the accuracy of the reference FVS is preserved provided that ε is chosen sufficiently small.

Corollary 7. *(Choice of threshold parameter)*
Let $\operatorname{diam}(V_{L,k}) \sim 2^{-L}$ *for all* $k \in I_L$. *If the discretization error of the reference FVS is bounded by*

$$\|\hat{\mathbf{u}}_L^n - \mathbf{v}_L^n\|_{1,L} \leq C\, 2^{-\alpha\, L}, \quad \alpha > 0,$$

where $\hat{\mathbf{u}}_L^n$ *denotes the average of the exact solution, then the accuracy is preserved by the adaptive scheme provided that*

$$\varepsilon \sim 2^{-(1+\alpha)\, L}.$$

Proof. Since for an explicit FVS the time step τ is restricted by the CFL condition, i.e., τ is proportional to the maximal diameter of the cells, here 2^{-L}, the assertion follows by Theorem 5. □

The usefulness of the above theorem crucially depends on the verification of the assumptions (A1) — (A5). First of all, we remark that for scalar conservation laws on $\Omega = \mathsf{R}^d$ there are several first order accurate schemes which are l^1–contractive, so–called monotone schemes, see e.g. [CM80]. Furthermore, we notice that (A2) is satisfied if the local flux evaluation is performed according to (4.13). In this case no error is introduced, i.e., $C_2 = 0$. In case of (4.14), (4.15) and (4.16), respectively, we introduce an error, but so far this error has not been verified to be bounded by ε. Concerning the consistent discretization of the initial data a natural choice is given by the approximation operator, i.e., $\overline{\mathbf{v}}_L^0 := \mathcal{A}_{\mathcal{D}^0_{L,\varepsilon}} \mathbf{v}_L^0$. In this case, (A5) holds provided the approximation error is uniformly bounded in the sense of (A1). We emphasize that the application of the approximation operator in general requires $\mathcal{O}(N_L)$ operations. In practice, a cheaper strategy is preferable, see Section 4.2.

In the following sections we now address to some extent the stability of the approximation operator and the reliability of the predictive set of significant details.

5.2 Stability of Approximation

In order to verify the stability of the approximation error it is convenient to investigate the convergence of the so–called *subdivision algorithms.* For this purpose, we rewrite the inverse local multiscale transformation (3.21) — (3.23) in full matrix–vector representation

$$\hat{\mathbf{u}}_{j+1} = \widetilde{\mathsf{G}}_{j,0}^T \, \hat{\mathbf{u}}_j + \sum_{e \in E^*} \widetilde{\mathsf{G}}_{j,e}^T \, \mathbf{d}_{j,e}.$$

Recursively applying this two–scale relation yields

$$\hat{\mathbf{u}}_L = \mathsf{G}_{0,0}^L \, \hat{\mathbf{u}}_0 + \sum_{j=0}^{L-1} \sum_{e \in E^*} \mathsf{G}_{j,e}^L \, \mathbf{d}_{j,e}, \tag{5.4}$$

where the matrix $\mathsf{G}_{j,e}^L \in \mathsf{R}^{N_L \times N_j}$ is defined by

$$\mathsf{G}_{j,e}^L := \widetilde{\mathsf{G}}_{L-1,0}^T \cdots \widetilde{\mathsf{G}}_{j+1,0}^T \, \widetilde{\mathsf{G}}_{j,e}^T \tag{5.5}$$

represents the *subdivision procedure.* Next we introduce the kth column of $\mathsf{G}_{j,e}^L$ which can be regarded as discrete basis vectors

$$\boldsymbol{\Psi}_{j,k,e}^L := \boldsymbol{\Psi}_{j,e}^L \, \mathbf{c}_{j,k} \tag{5.6}$$

by means of the Dirac vector $\mathbf{c}_{j,k} := (\delta_{k,r})_{r\in I_j}$, see [Dah94]. Then we can rewrite (5.4) as

$$\hat{\mathbf{u}}_L = \sum_{k\in I_0} \mathbf{\Psi}^L_{0,k,0}\, \hat{u}_{0,k} + \sum_{j=0}^{L-1} \sum_{e\in E^*} \sum_{k\in I_j} \mathbf{\Psi}^L_{j,k,e}\, d_{j,k,e}. \tag{5.7}$$

Here we only consider a scalar problem. In case of a system of conservation laws we can apply the scalar investigations componentwise. Consequently, the approximation error can be written in the form

$$\hat{\mathbf{u}}_L - \mathcal{A}_{\mathcal{D}}\, \hat{\mathbf{u}}_L = \sum_{(j,k,e)\notin\mathcal{D}} \mathbf{\Psi}^L_{j,k,e}\, d_{j,k,e}. \tag{5.8}$$

In order to control the threshold error we need the subdivision scheme to converge at least in the l^1–metric, i.e., the piecewise constant function $\psi^L_{j,k,e}$ defined by (5.9) converges to a function $\psi_{j,k,e}$ in L^1. It can be proved that the convergence of the subdivision scheme corresponds to the existence of a biorthogonal wavelet system. For general surveys on subdivision algorithms we refer to [Dyn92, CDM91] and to [Dau92, Coh00] for their relations to wavelets. So far, results are only available in case of uniform refinements on structured grids, see e.g. [CDF92, DKU99, CDKP00]. In case of curvilinear grids or arbitrary grids no stability results are available yet.

In the sequel, we employ some standard results from the theory of subdivision schemes which shall be summarized for the convenience of the reader. For this purpose we first introduce the notion of quasi–uniform grids.

Definition 9. *(Quasi–uniform Grid)*
Let $\mathcal{G}$ be a nested grid hierarchy. Then the grid on level j is called quasi–uniform *if there are constants $0 < c \le C$ such that*

$$c\,2^{-j} \le \operatorname{diam}(V_{j,k}) \le C\,2^{-j} \quad \text{and} \quad c\,2^{-jd} \le |V_{j,k}| \le C\,2^{-jd}$$

for all positions $k \in I_j$ and levels $j = 0,\ldots,L,$.

Then we obtain for the primal wavelets the following results.

Proposition 3.
Assume that the piecewise constant functions $\psi^L_{j,k,e}$ defined by

$$\psi^L_{j,k,e}(\mathbf{x}) := (\mathbf{\Psi}^L_{j,k,e})_r,\ \mathbf{x} \in V_{L,r},\ r \in I_L \tag{5.9}$$

converge uniformly in L towards a function $\psi_{j,k,e} \in L^\infty(\Omega)$ in the sup–norm. Then the limit functions (primal wavelets)[1] *satisfy the following properties:*

1.) Any function $u \in L^\infty(\Omega)$ can be uniquely expanded in a series of the primal wavelet basis, i.e.,

$$u = \sum_{k\in I_0} \langle u, \tilde{\varphi}_{0,k}\rangle_{L^2}\, \varphi_{0,k} + \sum_{j\in\mathbb{N}} \sum_{k\in I_j} \sum_{e\in E^*} \langle u, \tilde{\psi}_{j,k,e}\rangle_{L^2}\, \psi_{j,k,e}\ ;$$

[1] The functions $\psi_{j,k,0} = \varphi_{j,k}$ and $\tilde{\psi}_{j,k,0} = \tilde{\varphi}_{j,k}$ will also be referred to as wavelets.

2.) the primal wavelets satisfy the duality relation

$$\langle \psi_{j,k,e}, \tilde{\psi}_{j',k',e'} \rangle_{L^2} = \delta_{(j,k,e),(j',k',e')};$$

3.) the components of the discrete basis vectors coincide with the averages of the function $\psi_{j,k,e}$, i.e.,

$$\mathbf{\Psi}^L_{j,k,e} = (\langle \psi_{j,k,e}, \tilde{\varphi}_{L,r} \rangle)_{r \in I_L};$$

4.) the functions $\psi_{j,k,e}$ are uniformly bounded in the sup–norm, i.e., there exists a constant $C > 0$ independent of j, k and e such that

$$\|\psi_{j,k,e}\|_{L^\infty} < C;$$

5.) if the grid is quasi–uniform and the mask matrices $\widetilde{\mathsf{G}}_{j,e}$ are uniformly banded, then the functions $\psi_{j,k,e}$ are compactly supported and, in particular,

$$|\mathrm{supp}\, \psi_{j,k,e}| \le C\, 2^{-jd}.$$

A proof of the existence of the wavelet expansion can be found in [Dau92, Coh00], for instance. From the existence we conclude the biorthogonality property where we consider 1.) for $u = \psi_{j,k,e}$. Then the duality relation and applying (5.7) to $u = \psi_{j,k,e}$ implies that the coefficients of the discrete basis vectors coincide with the averages of the primal wavelets. Furthermore the uniform boundedness of the primal wavelets in the sup–norm is a consequence of the uniform convergence of the subdivision scheme. Finally, in [CDM91] it is proved that the primal wavelets are compactly supported.

In case of the box wavelets, see Section 2.3, the subdivision scheme has a simple representation. Therefore it is presented here for the convenience of the reader.

Example 4. (Box Wavelet)
We first observe that

$$\check{\mathsf{G}}^T_{j,e}\, \mathbf{c}_{j,k} = \sum_{r \in \mathcal{M}^0_{j,k}} \check{g}^{j,e}_{k,r}\, \mathbf{c}_{j+1,r} = \sum_{r \in \mathcal{M}^0_{j,k}} (a^{j,k}_{e,r} / a^{j,k}_{0,r})\, \mathbf{c}_{j+1,r}, \tag{5.10}$$

where we make use of (2.31). According to the definition of the subdivision scheme (5.5) we then deduce

$$\mathbf{\Psi}^L_{j,k,0} = \sum_{r \in \mathcal{M}^0_{j,k}} \mathsf{G}^L_{j+1,0}\, \mathbf{c}_{j+1,r} = \sum_{r \in \Sigma_{j,k,0}} \mathbf{c}_{L,r}, \tag{5.11}$$

where $\Sigma_{j,k,0}$ denotes the support of $\mathbf{\Psi}^L_{j,k,0}$, see also Section 4.1.2. By means of (5.5), (5.6) and (5.10) we further conclude

$$\mathbf{\Psi}^L_{j,k,e} = \mathsf{G}^L_{j+1,0}\, (\check{\mathsf{G}}^T_{j,e}\, \mathbf{c}_{j,k}) = \sum_{r \in \mathcal{M}^0_{j,k}} (a^{j,k}_{e,r} / a^{j,k}_{0,r})\, \mathbf{\Psi}^L_{j+1,r,0}. \tag{5.12}$$

Then the components of the discrete basis functions are determined by

$$\left(\boldsymbol{\Psi}^L_{j,k,e}\right)_l = \begin{cases} \sum_{r\in\mathcal{M}^0_{j,k}} a^{j,k}_{e,r}/a^{j,k}_{0,r} & , l \in \Sigma_{j,k,0} \\ 0 & , elsewhere \end{cases},$$

where we use (5.11) in (5.12) and use the definition of the Dirac vectors. Thus we obtain by the Hölder inequality

$$\|\boldsymbol{\Psi}^L_{j,k,e}\|^2_{l^\infty} \le \sum_{r\in\mathcal{M}^0_{j,k}} (a^{j,k}_{e,r})^2 \sum_{r\in\mathcal{M}^0_{j,k}} (a^{j,k}_{0,r})^{-2} = \|\mathbf{a}^{j,k}_e\|^2_2 \sum_{r\in\mathcal{M}^0_{j,k}} (|V_{j,k}|/|V_{j+1,r}|).$$

Since the vectors $\mathbf{a}^{j,k}_e$, $e \in E$, are assumed to be orthonormal, the right hand side can be estimated by a constant independent of j, k and e provided that the grids are quasi–uniform. Consequently, the discrete wavelets $\boldsymbol{\Psi}^L_{j,k,e}$ are uniformly bounded in the sup–norm.

We now introduce the piecewise constant functions $\psi^L_{j,k,e}$ according to (5.9) which can here be represented by

$$\psi^L_{j,k,e} = \sum_{r\in\mathcal{M}^0_{j,k}} (a^{j,k}_{e,r}/a^{j,k}_{0,r})\, \chi_{V_{j+1,r}}.$$

We notice that the functions $\psi^L_{j,k,e}$, $e \in E$, are independent of L, i.e., the subdivision scheme converges strongly towards $\psi_{j,k,e} := \psi^L_{j,k,e}$. Finally, we observe that the limit functions are uniformly bounded with respect to the sup–norm, because the relation $\|\psi^L_{j,k,e}\|_{L^\infty} = \|\boldsymbol{\Psi}^L_{j,k,e}\|_{l^\infty}$ holds.

From the above observations we now conclude the following stability result.

Theorem 6. *(Stability of approximation operator)*
Let $\varepsilon_j = 2^{(j-L)d}\varepsilon$ and $\varepsilon > 0$. If the computational domain is bounded, i.e. $|\Omega| < \infty$, and the grids are quasi–uniform, then the approximation operator is stable in the sense of

$$\|\hat{\mathbf{u}}_L - \mathcal{A}_\mathcal{D}\,\hat{\mathbf{u}}_L\|_{1,L} \le C\,\varepsilon$$

provided that the subdivision scheme converges strongly in L^∞.

Proof. According to (5.8) the approximation error can be estimated by

$$\|\hat{\mathbf{u}}_L - \mathcal{A}_\mathcal{D}\,\hat{\mathbf{u}}_L\|_{1,L} \le \sum_{(j,k,e)\notin\mathcal{D}} \|\boldsymbol{\Psi}^L_{j,k,e}\|_{1,L}\, |d_{j,k,e}| \le \sum_{(j,k,e)\notin\mathcal{D}} \|\boldsymbol{\Psi}^L_{j,k,e}\|_{1,L}\, \varepsilon_j.$$

Here we employ Definition 3.1 of the set $\mathcal{D}$ and not the scaled significant details according to (3.38), because a maximum–minimum principle holds for scalar conservation laws, see e.g. [CM80].

The norm of the discrete basis vectors can be represented as

$$\|\boldsymbol{\Psi}^L_{j,k,e}\|_{1,L} = \sum_{r\in\Sigma^L_{j,k,e}} |V_{L,r}|\,|(\boldsymbol{\Psi}^L_{j,k,e})_r|$$

where $\Sigma^L_{j,k,e} \subset I_L$ is the support of $\boldsymbol{\Psi}^L_{j,k,e}$. Next we conclude from Proposition 3

$$|(\boldsymbol{\Psi}^L_{j,k,e})_r| = |\langle \psi_{j,k,e}, \tilde{\varphi}_{L,r}\rangle| \le \|\psi_{j,k,e}\|_{L^\infty} \le C$$

and

$$\left|\bigcup\nolimits_{r\in\Sigma^L_{j,k,e}} V_{L,r}\right| = |\mathrm{supp}\,\psi_{j,k,e}| \le C\,2^{-jd}$$

Combining these two estimates we finally obtain

$$\|\hat{\mathbf{u}}_L - \mathcal{A}_{\mathcal{D}}\,\hat{\mathbf{u}}_L\|_{1,L} \le \overline{C} \sum_{(j,k,e)\notin\mathcal{D}} 2^{-jd}\,\varepsilon_j.$$

Since the grid is assumed to be quasi–uniform, we conclude that $N_j \sim 2^{jd}$ and, hence, the number of non–significant details is bounded by

$$\sum_{(j,k,e)\notin\mathcal{D}} 1 \le \sum_{(j,k,e)\notin\mathcal{D}_{L,0}} 1 = \sum_{j=0}^{L-1}\sum_{e\in E^*}\sum_{k\in I_j} 1 = N_L - N_0 \le C\,2^{Ld}.$$

This proves the assertion. □

We remark that the above threshold values coincide with Harten's choice in [Har95].

A similar result has been obtained in [Got98] where the uniform stability of the multiscale transformation related to the L^2–normalized counterparts of the modified box wavelets is employed. This is directly related to the existence of biorthogonal wavelets. In this setting we can apply well–known stability results corresponding to Cartesian grids. Finally, we point out that in the univariate dyadic case the L^2–normalized counterparts of the modified box wavelets according to Algorithm 1 coincide with the biorthogonal wavelets corresponding to the dual pair of B–splines of order 1 and dual generators of order 1 in [CDF92] ($\Omega = \mathsf{R}$) and [DKU99] ($\Omega = [a,b]$). In this case the stability of the multiscale transformation is guaranteed. By means of tensor products these can be extended to the multivariate dyadic case. However, the resulting wavelets do not coincide with those constructed in Algorithm 1.

5.3 Reliability of Prediction

The crucial assumption to be satisfied in Theorem 5 concerns the reliability of the prediction set $\tilde{\mathcal{D}}^{n+1}_{L,\boldsymbol{\varepsilon}}$. For this purpose we have to verify that the estimate

$$|d^{n+1}_{j',k',e'}| \le C\,\varepsilon_j \tag{5.13}$$

holds for $(j', k', e') \notin \tilde{\mathcal{D}}_{L,\varepsilon}^{n+1}$ where C is a constant independent of j', k' and e'. Note, that it suffices to bound the detail by a constant times the threshold value because otherwise we replace ε by ε/C. As we infer from the evolution equation for the details (4.9) the main task is to estimate the details of the flux balances.

In collaboration with A. Cohen et al. we have been able to prove that the estimate (5.13) holds provided that the prediction set is determined by the modified approach (4.21). The main problem arises from the nonlinearity of the flux balances. The objective is to estimate the flux balances by the details but they do not depend directly on the detail coefficients. To this end we interpret the details as finite differences. Then the finite differences of the flux balances can be estimated by finite differences of the averages where we make use of the chain rule. By means of an inverse estimate the finite differences of the averages can be estimated by details. For this purpose, we have to confine the discussion to a scalar nonlinear conservation law in one space dimension and a uniform discretization, see [CKMP01].

The proof consists of three basic steps:

- For any smooth function u the details of the corresponding wavelet expansion can be estimated by

$$|d_{j,k}| \leq C\, 2^{-sj}\, |u|_{C^s(\tilde{\Sigma}_{j,k})}.$$

  This estimate can be shown provided that $u \in C^s(\tilde{\Sigma}_{j,k})$, the size of the support $\tilde{\Sigma}_{j,k}$ of the modified box wavelet is of order $\mathcal{O}(2^{-j})$ and the wavelets are L^1-normalized, see [CKMP01]. A discrete counterpart is proved in Proposition 4 where we estimate the details by finite differences.

- The derivatives of any composite function $g \circ u$ of smooth functions g and u can be represented by

$$\frac{\partial^R\left(g(u(x))\right)}{\partial x^R} = \sum_{i=1}^{R} g^{(i)}(u(x)) \sum_{\substack{\mathbf{j}\in\{1,\dots,R-i+1\}^i \\ \mathbf{k}\in\{0,\dots,i\}^i \\ k_1 j_1+\dots k_i j_i = R}} c_{\mathbf{j},\mathbf{k},i} \prod_{l=1}^{i} \left(u^{(j_l)}(x)\right)^{k_l}$$

  successively applying the chain rule. This motivates the result of Proposition 5 where we derive a discrete counterpart by which the finite differences of the flux balances are estimated by the finite differences of the averages.

- Motivated by the standard inverse estimate

$$\|\Delta_h^r f\|_{L^p} \leq C\, \left(\min\{1, 2^j h\}\right)^r\, \|f\|_{L^p},$$

  with $\Delta_h^r f := \sum_{i=0}^{M} \binom{M}{i}(-1)^i f(\cdot + ih)$ we derive a discrete inverse estimate, see Proposition 6, from which we conclude that the finite differences can be estimated by the details.

In the sequel, the proof is outlined for the present setting. To this end, we confine the discussion to univariate dyadic grids, i.e.,

$$V_{j,k} = [2^{-j}\,k, 2^{-j}\,(k+1)],\;\; k \in I_j := \mathsf{Z},$$

where, in particular, the computational domain is assumed to be the real line, i.e., $\Omega = \mathsf{R}$. Since the assumption of shift–invariance simplifies the multiscale setting significantly we rewrite the basic relations according to Section 3.8. See also Sect. B.1.1 for details on shift–invariant spaces. We start with the two–scale relation for the averages

$$\hat{u}_{j,k} = 2^{-1}\,(\hat{u}_{j+1,2k} + \hat{u}_{j+1,2k+1}) = 2^{j-L} \sum_{i=0}^{2^{L-j}-1} \hat{u}_{L,2^{L-j}k+i}, \tag{5.14}$$

where we use (3.16) and $\check{\mathcal{M}}^0_{j,k} = \{2k, 2k+1\}$ as well as $|V_{j,k}| = 2^{-j}$. Moreover, we observe that the construction of the modified box wavelets according to Algorithm 1 with $\mathcal{L}^s_{j,k} = \{k-2s, \ldots, k+2s\}$, see (3.45), leads to coefficients $l^s_{j,k}$ which are independent of the level j and the position k. Note, that no boundary adaptation is necessary because $\Omega = \mathsf{R}$. Then the two–scale relation (3.17) — (3.20) for the details reads

$$d_{j,k} = \sum_{r=0}^{4s+1} m_r\, \hat{u}_{j+1,2(k-s)+r}, \tag{5.15}$$

where we have used that $\mathcal{M}^1_{j,k} = \{2(k-s), \ldots, 2(k-s)+1\}$. Since there is only one wavelet type in the univariate case, i.e., $E^* = \{1\}$, we omit the index e for the details.

In the following we heavily exploit the definition of a *finite difference of order* N which in the current setting reads

$$\Delta^N_K\, \hat{u}_{L,k} := \sum_{i=0}^{N} (-1)^i \binom{N}{i} \hat{u}_{L,k+iK} \tag{5.16}$$

with its stencil $S(N,K,k) := \{k+iK \,:\, i = 0, \ldots N\} \subset I_L$ depending on the order $N > 0$, the step size $K \in \mathsf{N}$ and the local position $k \in I_L$.

First of all, we verify that the details can be written as a linear combination of finite differences of order M.

Lemma 7. *(Characterization of details by finite differences)*
Let $k \in \mathsf{Z}$ and $j \in \{0, \ldots, L-1\}$. Then there are coefficients $\overline{m}_l$, $l = 0, \ldots, \overline{M} := 4s+1-M$, independent of k and j such that

$$d_{j,k} = \sum_{l=0}^{\overline{M}} \overline{m}_l\, \Delta^M_1\, \hat{u}_{j+1,2(k-s)+l} \tag{5.17}$$

provided that the modified box wavelet has M vanishing moments.

Proof. First of all we remark that the details vanish if the coefficients $\hat{u}_{l,r}$ represent the averages of a polynomial of degree $M-1$. In particular, this holds for the polynomials $p_i(x) := \sum_{m=0}^{i} a_{i,m} x^m$, $i = 0, \ldots, M-1$, where the coefficients $\mathbf{a}_i = (a_{i,m})_{m=0,\ldots,i}$ are determined by $\mathbf{a}_i^T := \mathbf{e}_i \, \mathsf{B}_i^{-1}$. Here $\mathbf{e}_i := (\delta_{i,m})_{m=0,\ldots,i}$ denotes the Dirac vector and $\mathsf{B}_i = (b^i_{m,r})_{m,r=0,\ldots,i}$ the lower triangular matrix defined by the elements

$$b^i_{m,r} := \begin{cases} \frac{1}{m+1} \binom{m+1}{r} 2^{-(j+1)m} & , 0 \le r \le m \\ 0 & , elsewhere \end{cases} .$$

Note, that the inverse B_i^{-1} exists because all diagonal elements are positive. For these polynomials the averages are determined by

$$\hat{p}_{i,j+1,l} = \frac{1}{|V_{j+1,l}|} \int_{V_{j+1,l}} p_i(x)\, dx = \mathbf{a}_i^T \, \mathsf{B}_i \, \mathbf{l}_i = l^i$$

with $\mathbf{l}_i = (l^m)_{m=0,\ldots,i}$. According to (5.15) we obtain for $k = s$

$$d_{j,s}(p_i) = \sum_{r=0}^{4s+1} m_r \, \hat{p}_{i,j+1,r} = \sum_{r=0}^{4s+1} m_r \, r^i = 0,$$

i.e., the vector of the mask coefficients $\mathbf{m} = (m_r)_{r=0,\ldots,4s+1}$ is orthogonal to the vectors $\mathbf{r}_i = (r^i)_{r=0,\ldots,4s+1}$, $i = 0, \ldots, M-1$. Since the vectors $\mathbf{r}_i$ are linearly independent, we conclude that $\mathbf{m}$ is orthogonal to the linear space $R_M := \mathrm{span}\{\mathbf{r}_0, \ldots, \mathbf{r}_{M-1}\}$.

Next we verify that the vectors $\mathbf{h}_i = (h_{l,i})_{l=0,\ldots,4s+1-M, i=0,\ldots,4s+1}$ defined by the elements

$$h_{l,i} := \begin{cases} (-1)^{i-l} \binom{M}{i-l} & , l \le i \le l+M \\ 0 & , elsewhere \end{cases}$$

form a basis for the complement space U_M of R^{4s+2}, i.e., $\mathsf{R}^{4s+2} = R_M \oplus U_M$. Obviously, these vectors are linearly independent. Therefore it remains to prove that the vectors $\mathbf{h}_l$, $l = 0, \ldots, 4s+1-M$, are orthogonal to the vectors $\mathbf{r}_i$, $i = 0, \ldots, M-1$. For this purpose we consider the inner product

$$\sum_{p=0}^{4s+1} h_{l,p} p^i = \sum_{p=l}^{l+M} (-1)^{p-l} \binom{M}{p-l} p^i = \sum_{r=0}^{i} \binom{i}{r} l^{i-r} A_{M,r}$$

with

$$A_{M,r} := \sum_{p=0}^{M} (-1)^p \binom{M}{p} p^r .$$

By induction over M we prove that $A_{M,r}$ vanishes for $r = 0, \ldots, M-1$. In this context, we employ the addition theorem for binomials.

Since $\mathbf{m}$ is orthogonal to R_M and $U_M = \text{span}\{\mathbf{h}_0, \ldots, \mathbf{h}_{4s+1-M}\}$ is orthogonal to R_M we conclude that $\mathbf{m} \in U_M$, i.e., there exist coefficients $\overline{m}_l$, $l = 0, \ldots, 4s+1-M$, such that

$$\mathbf{m} = \sum_{l=0}^{4s+1-M} \overline{m}_l \, \mathbf{h}_l .$$

Using this relation in (5.15) proves the assertion. □

From Lemma 7 we now deduce the result of the first step.

Proposition 4. *(Estimate of details by finite differences)*
Let $k \in \mathbb{Z}$, $j \in \{0, \ldots, L-1\}$ and $K := 2^{L-j-1}$. Furthermore, let M denote the number of vanishing moments of the modified box wavelet. Then the detail $d_{j,k}$ can be estimated by

$$|d_{j,k}| \leq C \sup\{|\Delta_K^M \hat{u}_{L,r}| \; ; \; r \in I_L \text{ s. t. } S(M,K,r) \subset \tilde{\Sigma}_{j,k}\}, \qquad (5.18)$$

where the constant C is independent of k and j.

Proof. First of all, we deduce from (5.16) and (5.14)

$$\Delta_1^M \hat{u}_{j,l} = 2^{j-L} \sum_{i=0}^{2^{L-j}-1} \Delta_{2^{L-j}}^M \hat{u}_{L,2^{L-j}l+i}.$$

By means of (5.15) and (5.17) we then conclude

$$d_{j,k} = \sum_{l=0}^{\overline{M}} \overline{m}_l \, 2^{j-L+1} \sum_{i=0}^{2^{L-j-1}-1} \Delta_{2^{L-j-1}}^M \hat{u}_{L,2^{L-j-1}(2(k-s)+l)+i}$$

and, consequently,

$$|d_{j,k}| \leq C \sup\{|\Delta_K^M \hat{u}_{L,K(2(k-s)+l)+i}| \; : \; i = 0, \ldots, K-1, \; l = 0, \ldots, \overline{M}\}.$$

Here the constant C is determined by the sum $C := \sum_{l=0}^{\overline{M}} |\overline{m}_l|$. According to the support $\tilde{\Sigma}_{j,k}$ of the modified box wavelets, see Section 4.1.2, only those finite differences are involved in the representation of the detail $d_{j,k}$ whose stencil $S(M,K,r)$ lies fully inside the support $\tilde{\Sigma}_{j,k}$. □

In the second step we apply the finite differences to the flux balances. The objective is to estimate these differences by those of the averages. Since the flux balances are in general determined by a nonlinear function this can not be done in a straightforward manner. However, it is possible to estimate the differences of the flux balances by lower order differences of the averages. To this end, we rewrite the numerical fluxes $F_{k,l}^{L,n}$ and the flux balances $f_{L,k}^n$, see (4.13) and (4.12), for the univariate case that is considered here. In agreement with Section 4.1.2 we assume that the stencil of the flux balances is determined by the set $\mathcal{N}_{L,k}^p$. Consequently, the numerical fluxes read

$$F_{k,k-1}^{L,n} = F\left(v_{L,k-p}^n, \ldots, v_{L,k+p-1}^n\right), \quad F_{k,k+1}^{L,n} = F\left(v_{L,k-p+1}^n, \ldots, v_{L,k+p}^n\right)$$

and thus

$$B_{L,k}^n := F_{k,k+1}^{L,n} - F_{k,k-1}^{L,n} = B\left(v_{L,k-p}^n, \ldots, v_{L,k+p}^n\right) \tag{5.19}$$

for a function $B : \mathsf{R}^{2p+1} \to \mathsf{R}$. In the following, the flux balance function B is assumed to be regular.

Assumption 3.
Let $D \subset \mathsf{R}^{2p+1}$ a bounded domain of admissible states. Then the flux balance function is assumed to be regular in the following sense:

1.) B is piecewise smooth, i.e., there are open subsets $D_i \subset D$, $i = 0, \ldots, K$, with $D = \bigcup_{i=0}^K \overline{D}_i$, such that $B \in C^R(D_i)$;

2.) B is Lipschitz-continuous on D;

3.) the derivatives of B can be extended continuously to the boundary ∂D_i such that

$$\sup_{\mathbf{v} \in \overline{D}_i} \frac{\partial^k B}{\partial^{k_0} v_0 \cdots \partial^{k_{2p}} v_{2p}}(\mathbf{v}) \leq C_k$$

for $k = \sum_{i=0}^{2p} k_i$, $k \in \{0, \ldots, R\}$.

Next we conclude that the adaptive FVS is uniformly bounded in the sup–norm provided that the reference FVS does not increase the sup–norm too much. In particular, the bound only depends on T and $\|u^0\|_{L^\infty}$.

Lemma 8. *(Boundedness of adaptive FVS in sup–norm)*
Assume that the following conditions hold:

(A6) the subdivision scheme converges uniformly in the sup–norm;

(A7) the local flux evaluation is based on the exact strategy, see (4.12) and (4.13);

(A8) the reference FVS satisfies

$$\|\mathcal{E}_L \mathbf{v}_L\|_{l^\infty} \leq (1 + C\,\tau)\, \|\mathbf{v}_L\|_{l^\infty};$$

(A9) the error of the initial data approximation can be estimated by

$$\|\overline{\mathbf{v}}_L^0 - \mathbf{v}_L^0\|_{l^\infty} \leq C\,\varepsilon/\tau \text{ and } \|\mathbf{v}_L^0 - \hat{\mathbf{u}}_L^0\|_{l^\infty} \leq C\,\varepsilon/\tau$$

where $\hat{\mathbf{u}}_L^0$ denotes the averages of the initial data;

(A10) the threshold values are determined by $\varepsilon_j = 2^{j-L}\,\varepsilon$ with $\varepsilon \sim 2^{-(1+\alpha)L}$ for some $\alpha > 0$;

(A11) the CFL condition holds on the finest resolution level, i.e., $\tau \sim 2^{-L}$.

Then the approximation $\overline{\mathbf{v}}_L^n$ corresponding to the adaptive FVS is uniformly bounded in the sup–norm, i.e.,

$$\|\overline{\mathbf{v}}_L^n\|_{l^\infty} \le C(T,u_0) \qquad \textit{for } n\,\tau \le T,$$

provided that in each time step the prediction set $\tilde{\mathcal{D}}_{L,\boldsymbol{\varepsilon}}^\nu$, $0 \le \nu \le T/\tau$, *satisfies the reliability property (4.19).*

Proof. In order to estimate the extrema by the time evolution we consider

$$\begin{aligned}\|\overline{\mathbf{v}}_L^n\|_{l^\infty} \le\; &\|\mathcal{A}_{\mathcal{D}_{L,\varepsilon}^{n-1}}\,\mathcal{E}_{\tilde{\mathcal{G}}_{L,\varepsilon}^{n-1}}\overline{\mathbf{v}}_L^{n-1} - \mathcal{E}_{\tilde{\mathcal{G}}_{L,\varepsilon}^{n-1}}\overline{\mathbf{v}}_L^{n-1}\|_{l^\infty} + \\ &\|\mathcal{E}_{\tilde{\mathcal{G}}_{L,\varepsilon}^{n-1}}\overline{\mathbf{v}}_L^{n-1} - \mathcal{E}_L\,\overline{\mathbf{v}}_L^{n-1}\|_{l^\infty} + \|\mathcal{E}_L\,\overline{\mathbf{v}}_L^{n-1}\|_{l^\infty}.\end{aligned} \tag{5.20}$$

According to Section 5.2 the first term can be estimated by

$$\left\|\mathcal{A}_{\mathcal{D}_{L,\varepsilon}^{n-1}}\,\mathcal{E}_{\tilde{\mathcal{G}}_{L,\varepsilon}^{n-1}}\overline{\mathbf{v}}_L^{n-1} - \mathcal{E}_{\tilde{\mathcal{G}}_{L,\varepsilon}^{n-1}}\overline{\mathbf{v}}_L^{n-1}\right\|_{l^\infty} \le \left\|\sum_{j=0}^{L-1}\;\sum_{(k,e)\in\mathcal{J}_{j,\varepsilon}^{n-1}} d_{j,k,e}^{n-1}\,\boldsymbol{\Psi}_{j,k,e}^L\right\|_{l^\infty}.$$

From assumption (A6) and Proposition 3 we conclude that the supports of the discrete basis vectors $\boldsymbol{\Psi}_{j,k,e}^L$ overlap only at a fixed number of positions independent of j, k and e. This implies

$$\left\|\sum_{(k,e)\in\mathcal{J}_{j,\varepsilon}^{n-1}} d_{j,k,e}^{n-1}\,\boldsymbol{\Psi}_{j,k,e}^L\right\|_{l^\infty} \le \sup_{(k,e)\in\mathcal{J}_{j,\varepsilon}^{n-1}} |d_{j,k,e}^{n-1}|\,\left\|\psi_{j,k,e}^L\right\|_{L^\infty} \le C\,\varepsilon_j, \tag{5.21}$$

where we employ that the prediction set is reliable in the sense of (4.19). Thus the first term can be estimated by

$$\left\|\mathcal{A}_{\mathcal{D}_{L,\varepsilon}^{n-1}}\,\mathcal{E}_{\tilde{\mathcal{G}}_{L,\varepsilon}^{n-1}}\overline{\mathbf{v}}_L^{n-1} - \mathcal{E}_{\tilde{\mathcal{G}}_{L,\varepsilon}^{n-1}}\overline{\mathbf{v}}_L^{n-1}\right\|_{l^\infty} \le \sum_{j=0}^{L-1}\varepsilon_j \le C\,\varepsilon.$$

The second term in (5.20) vanishes, because we use the exact local numerical flux evaluation. In this case, the evolution operators $\mathcal{E}_L$ and $\mathcal{E}_\mathcal{G}$ coincide. The third term can be estimated according to (A8) which results in the recursive estimate

$$\|\overline{\mathbf{v}}_L^n\|_{l^\infty} = \|\mathcal{E}_L\,\overline{\mathbf{v}}_L^{n-1}\|_{l^\infty} \le (1 + C\,\tau)\,\|\overline{\mathbf{v}}_L^{n-1}\|_{l^\infty} + C\,\varepsilon.$$

and, hence,

$$\|\overline{\mathbf{v}}_L^n\|_{l^\infty} \le (1 + C\,\tau)^n\,\|\overline{\mathbf{v}}_L^0\|_{l^\infty} + C\,n\,\varepsilon \le e^{C\,T}\,\|\overline{\mathbf{v}}_L^0\|_{l^\infty} + C\,T\,\varepsilon/\tau.$$

The initial data can be further estimated

$$\|\overline{\mathbf{v}}_L^0\|_{l^\infty} \le \|\hat{\mathbf{u}}_L^0\|_{l^\infty} + \|\overline{\mathbf{v}}_L^0 - \mathbf{v}_L^0\|_{l^\infty} + \|\mathbf{v}_L^0 - \hat{\mathbf{u}}_L^0\|_{l^\infty} \le \|u_0\|_{L^\infty} + 2\,C\,\varepsilon/\tau.$$

From assumption (A10) and (A11) we conclude that the ratio ε/τ is small in comparison to $\|u_0\|_{L^\infty}$. Consequently, we can estimate the supremum of $\overline{\mathbf{v}}_L^n$ by a constant only depending on T and the supremum of the initial data u_0. □

So far we have not verified the reliability property (4.19) for the prediction set $\tilde{\mathcal{D}}_{L,\varepsilon}^{n+1}$. However, we may assume that (4.19) holds for all previous time steps $\nu = 0, \ldots, n$ in order to check the reliability of the prediction set for the new time step. As we will see in the proof of Proposition 5, it will be sufficient to know that the data of the old time step are uniformly bounded by the constant $C(T, u_0)$.

Proposition 5. *(Finite differences for composite functions)*
Let the assumptions of Lemma 8 hold and assume that the flux balance function B satisfies Assumption 3. Introducing

$$D_N(\overline{\mathbf{v}}_L^n, K, \tilde{\Sigma}_{j,k}) := \sup \left\{ |\Delta_K^N \overline{v}_{L,k'}^n| \; ; \; S(N,K,k') \subset \tilde{\Sigma}_{j,k} \right\} \quad \textit{and}$$

$$I(R) := \left\{ (\mathbf{j},\mathbf{k}) \; ; \; \mathbf{j} \in \{1,\ldots,R\}^R, \; \mathbf{k} \in \{0,\ldots,R\}^R, \; \sum_{l=1}^{R} j_l \, k_l = R \right\},$$

we obtain

$$D_R(\mathsf{B}_L^n, K, \tilde{\Sigma}_{j,k}) \leq C \sup \left\{ \prod_{l=1}^{R} (D_{j_l}(\overline{\mathbf{v}}_L^n, K, \tilde{\Sigma}_{j,k}^-))^{k_l} \; ; \; (\mathbf{j},\mathbf{k}) \in I(R) \right\}. \tag{5.22}$$

Proof. We want to estimate the finite differences $\Delta_K^R B_{L,k}^n$ by means of finite differences of possibly lower order. To this end, we first introduce Lagrange polynomials $p_i \in \mathsf{P}_R$, $i = 0, \ldots, 2p$, of degree R defined by the interpolation conditions

$$p_i(m) = \overline{v}_{L,k-p+i+mK}^n, \quad m = 0, \ldots, R, \tag{5.23}$$

and the composite function $G : \mathsf{R} \to \mathsf{R}$

$$G(x) := B(p_0(x), \ldots, p_{2p}(x)) \equiv B(\mathsf{P}(x)).$$

The definition of the finite differences implies

$$\Delta_K^R B_{L,k}^n = \sum_{m=0}^{R} (-1)^m \binom{R}{m} G(m) =: \Delta_1^R G(0)$$

and thus we deduce from standard results for finite differences, see e.g. [SB80, DL93],

$$|\Delta_K^R B_{L,k}^n| \leq C \sup_{x \in [0,R]} \left| G^{(R)}(x) \right|. \tag{5.24}$$

Furthermore the R–th derivative of the composite function can be represented by

$$G^{(R)}(x) = \sum_{k=1}^{R} \sum_{\mathbf{i}\in\{0,2p\}^k} \frac{\partial^k B}{\partial p_{i_1} \cdots \partial p_{i_k}}(\mathsf{P}(x)) \sum_{\substack{\mathbf{j}\in\{1,\dots,R-k+1\}^k \\ j_1+\dots j_k=R}} c_{\mathbf{j},\mathbf{i},k} \prod_{l=1}^{k} p_{i_l}^{(j_l)}(x),$$

where we employ the smoothness of the flux balance function B according to Assumption 3. When $\mathsf{P}(x) \in \partial D_i$, we consider the onesided continuous extensions of the derivatives. This representation can be derived successively applying the chain rule for differentiation. Then we can estimate the R–th derivative by

$$\sup_{x\in[0,R]} \left|G^{(R)}(x)\right| \le C \sup_{x\in[0,R]} \left\{ \prod_{l=1}^{R} |p_{i_l}^{(j_l)}(x)|^{k_l} \ ; \ (\mathbf{j},\mathbf{k}) \in I(R) \right\}, \qquad (5.25)$$

where the constant C depends on the coefficients $c_{\mathbf{j},\mathbf{i},k}$ and R, respectively, and the bounds

$$\sup_{x\in[0,R]} \left\{ \frac{\partial^k B}{\partial p_{i_1} \cdots \partial p_{i_k}}(\mathsf{P}(x)) \ : \ \mathbf{i} \in \{0,\dots,2p\}^k \right\}.$$

From the Lagrangian representation of the polynomials p_i we conclude that there exists a uniform bound such that

$$\sup_{x\in[0,R]} |p_i(x)| \le C \, \|\overline{\mathbf{v}}_L^n\|_{l^\infty}, \quad i = 0,\dots,2p.$$

According to the assumptions and Lemma 8 we know that the reference FVS is uniformly bounded in the sup–norm. Hence, the bounds C_k only depend on T and $\|u_0\|_{L^\infty}$. We now consider the Newton representation of the interpolation polynomials p_i, i.e.,

$$p_i(x) = \sum_{\nu=0}^{R} p_i[0,\dots,\nu] \prod_{l=0}^{\nu-1} (x-l) = \sum_{\nu=0}^{R} \frac{1}{\nu!} \Delta_1^\nu \, p_i(0) \prod_{l=0}^{\nu-1} (x-l), \qquad (5.26)$$

where $p_i[0,\dots,\nu]$ denotes the ν–th divided difference, see e.g. [SB80, DL93]. By means of induction and using the addition theorem for binomial coefficients we notice that

$$\Delta_1^{\nu+j} \, p_i(0) = \sum_{l=0}^{\nu} \binom{\nu}{l} (-1)^l \, \Delta_1^j \, p_i(l), \quad \nu \ge 0.$$

Incorporating this relation in (5.26) we deduce

$$\sup_{x\in[0,R]} \left|p_i^{(j)}(x)\right| \le C \max_{l=0,\dots,R-j} |\Delta_1^j \, p_i(l)|, \quad j \le R, \qquad (5.27)$$

where C only depends on R. Note, that the terms for $\nu = 0,\dots,j-1$ in (5.26) give no contribution to the jth derivative. Furthermore the definition of the finite differences and the interpolation conditions (5.23) yield

$$\Delta_1^j\, p_i(l) = \Delta_K^j\, \overline{v}_{L,k-p+i+lK}^n. \tag{5.28}$$

Combining (5.24), (5.25), (5.27) and (5.28) we obtain

$$|\Delta_K^R\, B_{L,k}^n| \le C\ \sup\{\prod_{l=1}^R |\Delta_K^{j_l}\, \overline{v}_{L,k-p+i_l+\nu_l K}^n|^{k_l}\ ;\ (\mathbf{j},\mathbf{k}) \in I(R),$$
$$(\mathbf{i},\nu) \in I(R,p,\mathbf{j})\},$$

where $I(R,p,\mathbf{j}) := \left\{(\mathbf{i},\nu)\ ;\ \mathbf{i} \in \{0,\ldots,2p\}^R,\ \nu_l \in \{1,\ldots,R-j_l\}^R\right\}$. Note, that the bound C only depends on R and C_k. We now estimate the finite differences on the right hand side by

$$|\Delta_K^{j_l}\, \overline{v}_{L,k-p+i_l+\nu_l K}^n| \le D_{j_l}(\overline{\mathbf{v}}_L^n, K, \tilde{\Sigma}_{j,k}^-),$$

where we take into account that the set $\tilde{\Sigma}_{j,k}^-$ is inflated by the stencil of the numerical fluxes specified by the parameter p. Therefore we need $\tilde{\Sigma}_{j,k}^-$ instead of $\tilde{\Sigma}_{j,k}$. Finally we obtain the assertion taking the supremum over all k' such that $S(R,K,k') \subset \tilde{\Sigma}_{j,k}^-$. □

Next we derive a converse result to Proposition 3, i.e., we want to estimate the finite differences $\Delta_K^M\, \hat{u}_{L,k}$ by certain details $d_{j,r}$. For this purpose we employ an inverse (Bernstein–type) estimate which is a standard tool in approximation theory.

Lemma 9. *(Inverse estimate)*
Let $1 \le p, p' \le \infty$ such that $1/p + 1/p' = 1$. Assume that the scaling functions $\varphi_{j,k}$ and $\tilde{\varphi}_{j,k}$ are refinable functions, i.e., there are functions φ and $\tilde{\varphi}$ such that $\varphi_{j,k} = 2^{j\,d/2}\,\varphi(2^j\cdot - k)$ and $\tilde{\varphi}_{j,k} = 2^{j\,d/2}\,\tilde{\varphi}(2^j\cdot - k)$, respectively. If $\varphi \in W^{r,p}(\mathsf{R})$ and $\tilde{\varphi} \in L^{p'}(\mathsf{R})$, then for any function $f \in S_j = \mathrm{span}\{\varphi_{j,k}\ ;\ k \in \mathsf{Z}\}$

$$\|\Delta_h^r f\|_{L^p} \le C\ \left(\min\{1, 2^j h\}\right)^r\ \|f\|_{L^p} \tag{5.29}$$

holds, where the constant C is independent of j.

In case of $h \ge 2^{-j}$ the result can immediately be inferred from the definition of the finite differences for *any* $f \in L^p$. For the complementary case the smoothness of the approximation spaces S_j has to be exploited. For a detailed proof see [Coh00], Chapt. III.

Note, that we can apply in particular this result to the primal wavelets $\psi_{j,k} \in W_j \subset S_{j+1}$ we are considering here. Since the real axis is discretized by a uniform dyadic grid, the convergence of the subdivision scheme, see Section 5.2, ensures that the primal wavelet can be represented by the shifts and translates of a scaling function.

By means of the above lemma we are now able to prove the following inverse estimate.

Proposition 6. *(Discrete inverse estimate)*
Let $K = 2^{L-j'-1}$. Assume that the subdivision scheme converges uniformly in the sup–norm and the corresponding primal wavelets $\psi_{j,k}$ are in C^r. For $N > 0$ we obtain

$$|\Delta_K^N \hat{u}_{L,k'}| \le C \sum_{j=-1}^{L-1} 2^{-\min\{N,r\}(j'-j-1)_+} \sup\{|d_{j,k}| \; ; \; \Sigma_{j,k} \cap S(N,K,k') \neq \emptyset\}, \tag{5.30}$$

where $d_{-1,k} := \hat{u}_{0,k}$, $I_{-1} := I_0$, $\Sigma_{-1,k} := \Sigma_{0,k,0}$ and $(j'-j-1)_+ := \max\{j'-j-1, 0\}$.

Proof. The vector $\hat{\mathbf{u}}_L$ can be expanded in a series of the discrete wavelet basis according to (5.7), i.e.,

$$\hat{\mathbf{u}}_L = \sum_{j=-1}^{L-1} \sum_{k \in I_j} d_{j,k} \, \boldsymbol{\Psi}_{j,k}^L,$$

where we use the convention $\boldsymbol{\Psi}_{-1,k}^L \equiv \boldsymbol{\Psi}_{0,k,0}^L$ and $\boldsymbol{\Psi}_{j,k}^L \equiv \boldsymbol{\Psi}_{j,k,1}^L$, $j = 0, \ldots, L-1$, see Section 5.2. From this relation we deduce

$$|\Delta_K^N \hat{u}_{L,k'}| \le \sum_{j=-1}^{L-1} \sum_{k \in I_j} |d_{j,k} \, \Delta_K^N (\boldsymbol{\Psi}_{j,k}^L)_{k'}|.$$

Since the differences on the right hand side vanish if the supports $\Sigma_{j,k}$ and $S(N,K,k')$ do not overlap, the summation is only carried over those (j,k) such that $\Sigma_{j,k} \cap S(N,K,k') \neq \emptyset$. Furthermore, we know that the supports of the discrete wavelets $\boldsymbol{\Psi}_{j,k}^L$ overlap only at a fixed number of positions independent of j and k, see also the proof of Lemma 8. Obviously, this also holds for the differences $\Delta_K^N (\boldsymbol{\Psi}_{j,k}^L)_{k'}$. Similar to (5.21) we then obtain

$$|\Delta_K^N \hat{u}_{L,k'}| \le \sum_{j=-1}^{L-1} \sup\{|d_{j,k}| \, \|\Delta_K^N (\boldsymbol{\Psi}_{j,k}^L)_{k'}\|_{l^\infty} \; ; \; \Sigma_{j,k} \cap S(N,K,k') \neq \emptyset\}. \tag{5.31}$$

It remains to estimate the finite differences of the discrete wavelets in the sup–norm. To this end, we first recall that the discrete wavelet $\boldsymbol{\Psi}_{j,k}^L$ coincides with the cell averages of the continuous wavelet $\psi_{j,k}$, see Proposition 3, assertion 3. In particular, we obtain for $h = 2^{-L}K$

$$(\boldsymbol{\Psi}_{j,k}^L)_{k'+iK} = 2^L \int_{2^{-L}(k'+iK)}^{2^{-L}(k'+1+iK)} \psi_{j,k}(x)dx = 2^L \int_{2^{-L}k'}^{2^{-L}(k'+1)} \psi_{j,k}(x+ih)dx$$

and, hence,

$$\Delta_K^N (\boldsymbol{\Psi}_{j,k}^L)_{k'} = 2^L \int_{2^{-L}k'}^{2^{-L}(k'+1)} \Delta_h^N \psi_{j,k}(x)dx.$$

Therefore we can estimate the finite differences of the discrete wavelets by those for the continuous wavelets. We are now able to employ the inverse estimate (5.29) for $p = \infty$. In case of $N \le r$ it can be directly applied to $f = \psi_{j,k} \in S_{j+1}$. For $r < N$ we first observe by the recursive representation of the finite differences that

$$\Delta_h^N \psi_{j,k}(x) = \Delta_h^{N-r} (\Delta_h^r \psi_{j,k}(x)).$$

Finally, we end up with

$$\|\Delta_h^N \psi_{j,k}\|_{L^\infty} \le C \, \|\psi_{j,k}\|_{L^\infty} \, \left(\min\{1, h2^{j+1}\}\right)^{\min\{N,r\}}.$$

Combining this with (5.31) leads to (5.30), because the primal wavelets are uniformly bounded, see Proposition 3. □

In the last step we estimate the powers of the low order finite differences in Proposition 5 by the threshold values ε_j. For this purpose we exploit Proposition 6 where the finite differences are estimated by the details. In addition, we incorporate the definition of the prediction sets $\tilde{\mathcal{D}}_{L,\varepsilon}^{n+1}$, see (4.21). Note, that in the scalar case the norm $\|\cdot\|_{\infty,*}$ is replaced by the absolute value $|\cdot|$ in (4.20). However, we have to specify first the choice of the parameter σ in (4.20). To this end, we make the following assumption.

Assumption 4.
Assume that the primal wavelets have C^r Hölder smoothness, i.e., $\psi_{j,k} \in C^r$, and the dual wavelets have M vanishing moments. Then we choose some σ such that

$$1 < \sigma < r + 1 \tag{5.32}$$

and fix the parameters R and $\beta > 0$ such that

$$R - 1 < r \le R, \tag{5.33}$$

$$1 + \beta < \sigma < 1 + R - \beta. \tag{5.34}$$

Note, that the smoothness parameter r is bounded by the number of vanishing moments M of the dual wavelets $\tilde{\psi}_{j,k}$, i.e., $r < M$, and thus $\sigma < M + 1$ and $R \le M$.

Proposition 7.
Let the assumptions of Lemma 8 and Proposition 6 as well as Assumption 4 hold. Let $(j', k') \notin \tilde{\mathcal{D}}_{L,\varepsilon}^{n+1}$, $K := 2^{L-j'-1}$, $N > 0$ and k such that $S(N, K, k) \subset \tilde{\Sigma}_{j',k'}^{-}$. Then we get the estimate

$$|\Delta_K^N \overline{v}_{L,k}^n| \le C \, \varepsilon_{j'}^{\min\{N/R,1\}}, \tag{5.35}$$

where the threshold values are given by $\varepsilon_j = 2^{j-L} \varepsilon$. In particular, if $N < R$ then the constant C depends on T and u_0.

Proof. Since $S(N,K,k) \subset \tilde{\Sigma}^-_{j',k'}$, we infer from (5.30) the estimate

$$|\Delta_K^N \overline{v}^n_{L,k}| \le C \sum_{j=-1}^{L-1} 2^{-\min\{N,r\}\,(j'-j-1)_+} \sup\{|d^n_{j,l}|\ ;\ \Sigma_{j,l} \cap \tilde{\Sigma}^-_{j',k'} \neq \emptyset\}. \tag{5.36}$$

Again we notice that the supremum has only to be taken over all indices $(j,l) \in \mathcal{D}^n_{L,\varepsilon}$ such that $\Sigma_{j,l} \cap \tilde{\Sigma}^-_{j',k'} \neq \emptyset$, since otherwise $d^n_{j,l} = 0$. The details and the coarse–scale averages $\overline{v}^n_{0,k} = d^n_{-1,k}$ corresponding to these indices are bounded by

$$|d^n_{j,l}| \le \varepsilon_j\, 2^{(\nu(j,l)+1)\sigma} \le \varepsilon_j\, 2^{(j'-j)\sigma} = 2^{j-L+(j'-j)\sigma}\, \varepsilon, \tag{5.37}$$

because $(j',k') \notin \tilde{\mathcal{D}}^{n+1}_{L,\varepsilon}$ implies $j' > j + \nu(j,l)$ according to (4.21).

We first consider $N \ge R$. Since R is chosen such that (5.33) holds, this implies $\min\{N,r\} = r$. Then we can further estimate the finite differences where we combine (5.36) and (5.37)

$$|\Delta_K^N \overline{v}^n_{L,k}| \le C\,\varepsilon\, 2^{-L} \sum_{j=-1}^{L-1} 2^{-r\,(j'-j-1)_+}\, 2^{\sigma\,(j'-j)+j} = C\,\varepsilon\, 2^{-L}\,(\Sigma_1 + \Sigma_2)$$

with

$$\Sigma_1 := 2^{(\sigma-r)j'+r} \sum_{j=-1}^{j'-2} 2^{j(r-\sigma+1)}, \quad \Sigma_2 := 2^{\sigma j'} \sum_{j=j'-1}^{L-1} 2^{j(1-\sigma)}.$$

Since (5.32) holds, we can estimate Σ_1 and Σ_2 by $C\,2^{j'}$ which proves (5.35) in this case.

Next we consider $N < R$. Then (5.33) implies $\min\{N,r\} = N$. First of all, we conclude from (5.15) and (5.14) that

$$|d^n_{j,l}| \le C\,\|\overline{\mathbf{v}}^n_L\|_{l^\infty}.$$

Note, that this also holds for $d^n_{-1,0} = \overline{v}^n_{0,l}$. Combining this with (5.37) we infer from Lemma 8

$$|d^n_{j,l}|^p = |d^n_{j,l}|^{p-1}\, |d^n_{j,l}| \le C\, 2^{j-L+(j'-j)\sigma}\, \varepsilon \tag{5.38}$$

provided that $p > 1$. We notice that C is proportional to $(C(T,u_0))^{p-1}$. Now we put $p := R/N$. Then we can estimate the pth power of the finite differences where we use Hölder's inequality

$$|\Delta_K^N \overline{v}^n_{L,k}|^p \le C \sum_{j=-1}^{L-1} a_j\, b_j \le C \sum_{j=-1}^{L-1} (a_j)^p \left(\sum_{j=-1}^{L-1} (b_j)^{p/(p-1)} \right)^{p-1}$$

with

$$a_j := 2^{-M\,(j'-j-1)_+}\,2^{\beta\,|j'-j-1|\sigma/p}\,s_j, \quad s_j := \sup\{|d^n_{j,l}|\ ;\ \Sigma_{j,l}\cap\tilde{\Sigma}^-_{j',k'} \neq \emptyset\}$$

$$b_j := 2^{-\beta\,|j'-j-1|\sigma/p}.$$

The second sum in the right hand side is bounded by a constant, since $2^{-\beta\sigma/(p-1)} < 1$. In order to estimate the first sum, we first notice that the pth power of the supremum s_j is bounded by

$$s_j^p \leq \sup\{|d^n_{j,l}|^p\ ;\ \Sigma_{j,l}\cap\tilde{\Sigma}^-_{j',k'} \neq \emptyset\} \leq C\,2^{j-L+(j'-j)\sigma}\,\varepsilon,$$

where we employ (5.38). Then we obtain

$$\begin{aligned}\sum_{j=-1}^{L-1}(a_j)^p &\leq C\,\varepsilon\,2^{-L}\sum_{j=-1}^{L-1} 2^{-R\,(j'-j-1)_+}\,2^{\beta\,|j'-j-1|}\,2^{j+(j'-j)\sigma}\\ &\leq C\,\varepsilon\,2^{-L}\,(\Sigma_1+\Sigma_2)\end{aligned}$$

with

$$\Sigma_1 := 2^{(-R+\beta+\sigma)j'+R-\beta}\sum_{j=-1}^{j'-2} 2^{j(R-\beta-\sigma+1)},$$

$$\Sigma_2 := 2^{(\sigma-\beta)j'+R}\sum_{j=j'-1}^{L-1} 2^{j(\beta-\sigma+1)}.$$

Both terms are bounded by $C\,2^{j'}$, since β is chosen such that (5.34) holds. This proves the assertion. □

Now we are able to verify the reliability property (4.19) for the prediction set (4.21). For this purpose, we combine the results of the Propositions 4, 5 and 7.

Theorem 7. *(Reliability, [CKMP01])*
Consider the univariate shift–invariant setting and let the Assumptions 3, 4 as well as the assumptions (A6) — (A11) of Lemma 8 hold. Then the prediction set defined by (4.21) fulfills the reliability property (4.19).

Proof. Let $(j',k') \notin \tilde{\mathcal{D}}^{n+1}_{L,\varepsilon}$. Then we have to verify that the corresponding detail on the new time level is not significant, i.e., (5.13) holds. In the first step we estimate the detail by the supremum of certain finite differences of order M according to Proposition 4, i.e.,

$$|d^{n+1}_{j',k'}| \leq C\,\sup\{|\Delta^M_K\,\overline{v}^{n+1}_{L,r}|\ ;\ r\in I_L \text{ s. t. } S(M,K,r)\subset\tilde{\Sigma}_{j',k'}\} \qquad (5.39)$$

with $K := 2^{L-j'-1}$ and M the number of vanishing moments of the dual wavelets. Next we apply the evolution equation (4.1) to the averages $\overline{v}^{n+1}_{L,r}$

where we replace the flux balances $\Delta F^n_{L,k}$ by $B^n_{L,k}$ according to (5.19). Then the finite differences in (5.39) can be estimated by

$$|\Delta_K^M \overline{v}_{L,r}^{n+1}| \le |\Delta_K^M \overline{v}_{L,r}^n| + \lambda_{L,r}\, |\Delta_K^M B_{L,r}^n|. \tag{5.40}$$

Since $R \le M$, Proposition 7 implies

$$|\Delta_K^M \overline{v}_{L,r}^n| \le C\, \varepsilon_{j'}. \tag{5.41}$$

From the CFL constraint we furthermore deduce

$$\lambda_{L,r} = 2^L \tau \le C\, \left(\max_{|u|\le C(T,u_0)} |f'(u)|\right)^{-1} =: C(T,u_0,f'). \tag{5.42}$$

Combining (5.39), (5.40), (5.41) and (5.42) we obtain

$$|d_{j',k'}^{n+1}| \le C\, \varepsilon_{j'} + C(T,u_0,f')\, D_M(\mathsf{B}_L^n, K, \tilde{\Sigma}_{j',k'}).$$

Here we employ the definition of $D_M(\mathsf{B}_L^n, K, \tilde{\Sigma}_{j',k'})$ introduced in Proposition 5. Hence it remains to verify that the differences of the flux balances are bounded by $C\,\varepsilon_{j'}$. To this end, recall the meaning of R from (5.33) and note that

$$|\Delta_K^M B_{L,r}^n| = |\Delta_K^{M-R}\, \Delta_K^R B_{L,r}^n| \le C\, |\Delta_K^R B_{L,r}^n|.$$

Hence we infer that

$$D_M(\mathsf{B}_L^n, K, \tilde{\Sigma}_{j',k'}) \le C\, D_R(\mathsf{B}_L^n, K, \tilde{\Sigma}_{j',k'}). \tag{5.43}$$

In order to estimate the right hand side we apply (5.22) to obtain

$$D_R(\mathsf{B}_L^n, K, \tilde{\Sigma}_{j',k'}) \le C \sup \left\{ \prod_{l=1}^R (D_{j_l}(\overline{\mathbf{v}}_L^n, K, \tilde{\Sigma}_{j',k'}^-))^{k_l} \;;\; (\mathbf{j},\mathbf{k}) \in I(R) \right\}. \tag{5.44}$$

For $(\mathbf{j},\mathbf{k}) \in I(R)$ and $r \in I_L$ such that $S(j_l, K, r) \subset \tilde{\Sigma}_{j',k'}^-$, $1 \le l \le R$, we conclude from Proposition 7

$$|\Delta_K^{j_l} \overline{v}_{L,r}^n| \le C\, \varepsilon_{j'}^{\min\{1, j_l/R\}}.$$

Thus, the supremum can also be estimated by this bound, i.e.,

$$D_{j_l}(\overline{\mathbf{v}}_L^n, K, \tilde{\Sigma}_{j',k'}^-) \le C\, \varepsilon_{j'}^{\min\{1, j_l/R\}}$$

and we obtain

$$\prod_{l=1}^R (D_{j_l}(\overline{\mathbf{v}}_L^n, K, \tilde{\Sigma}_{j',k'}^-))^{k_l} \le C \prod_{l=1}^R \varepsilon_{j'}^{\min\{k_l, j_l k_l/R\}} = C\, \varepsilon_{j'}^{\sum_{l=1}^R j_l k_l/R}.$$

Since $(\mathbf{j},\mathbf{k}) \in I(R)$, we conclude from Assumption 4

$$\sum_{l=1}^R \frac{j_l k_l}{R} = \frac{R}{R} = 1.$$

Finally, the right hand side of (5.44) can be estimated by

$$D_R(\mathsf{B}_L^n, K, \tilde{\Sigma}_{j',k'}) \le C\, \varepsilon_{j'}.$$

Combining this with (5.43) proves the assertion. □

6 Data Structures and Memory Management

In the context of adaptive schemes the design of *data structures* and an appropriate *memory management* have a significant influence on the performance of the computation. In the following we will discuss the design of data structures where we deduce some fundamental *design criteria* from *algorithmic requirements*, see Sect. 6.1. Here the concept of *hashing* turns out to be an efficient tool which fulfills these criteria. The basic ideas of *hash tables*, *hash functions* and *collision strategies* are summarized in Sect. 6.2, although they can be found in standard text books, see e.g. [Str97, OW96]. Nevertheless we believe it to be convenient for the reader to repeat some details, since we add some special features to the standard setting which are needed for the application at hand. Finally, in Sect. 6.3 we present the data types by which we realize the adaptive code. These are derived from the template library `igpm_t_lib` developed at the Institut für Geometrie und Praktische Mathematik, RWTH Aachen, see [MV00].

6.1 Algorithmic Requirements and Design Criteria

In the previous chapters we developed an adaptive scheme with an optimal complexity in the sense that the number of operations is proportional to the number of unknowns, i.e., $\#\mathcal{D}_{L,\varepsilon}$ and $\#\mathcal{G}_{L,\varepsilon}$, respectively. We are now concerned with the design of an optimal code in the sense that the memory requirements and the CPU time are proportional to the complexity of the adaptive algorithm. Here it turns out that the choice of *data structures* and *memory management* has a significant influence on the performance of the computation. Our experience shows that the design of appropriate data structures crucially depends on the underlying adaptive algorithm, i.e., the data structures have to be adapted to the algorithmic requirements and should not be designed independently. Therefore we first have to determine these requirements. To this end, we consider the Algorithms 3 and 6 by which the local (inverse) multiscale transformations are performed, since these form the backbone of the adaptive scheme. From these we deduce the following demands:

- The local transformations are performed level by level. Therefore data structures by which the local averages and the significant details are stored have to maintain the *level information.*
- For each level we have to collect the indices of cells to be refined or coarsened. Therefore we need a data structure for index sets.
- For the computation of the local (inverse) two–scale transformation (3.16) — (3.20) and (3.21) — (3.23), respectively, we have to determine the *supports* of the mask matrices $\check{\mathsf{M}}_{j,0}$, $\check{\mathsf{G}}_{j,0}$ and $\mathsf{L}_{j,e}$. The summation is then performed on a column of these matrices. Obviously, we never access a single element of the matrices but a *column* corresponding to the non–vanishing entries indicated by the support. This data group has to be stored as *one* element by a data structure.
- While performing the local transformations we have to check whether a matrix column or a local average as well as a significant detail *exists*, i.e., whether this information has already been computed and has been stored.
- The data provided by the local transformations have to be *inserted* or have to be *deleted* from a current data structure.
- If we access details $d_{j,k,e}$, we always consider *all* wavelet types indicated by $e \in E^*$.
- The algorithm is independent of the spatial dimension d and the number of conservation laws m. Therefore the data structures have to be designed *flexibly* with respect to this parameters.

From these algorithmic requirements we deduce the design criteria for an appropriate data structure for our sparse data distribution. This data structure should provide

- *fast random data access*, e.g. check whether an element already exists,
- *fast inserting and deleting* of elements, i.e., copying and sorting of elements within the data structure should be avoided,
- *fast dynamic memory allocation and extension*, since the memory requirement is not a priori known but can only be approximately predicted and
- *support group information*, i.e., connections of data corresponding to a common level, should be maintained.

By means of these design criteria the template library `igpm_t_lib` has been developed, see [MV00]. This library provides template classes for

- a *vector* with an arbitrary but fixed dimension, see `tvector_n`,
- an *index* composed of a multiindex or, additionally, indicated by a level and blocks of multiindices representing the supports, see `tmulti-index`, `tlevelmultiindex` and `tmultirange`,

– *hash maps* by which elements can be stored in an unordered way providing fast access, see `thashtab`, `thashtab_linked` and `thashtab_linked_one`.

The fundamental data structures used in the implementation of the adaptive scheme are various hash maps which optionally support grouping of the data by means of level information.

6.2 Hashing

In the context of adaptive schemes it is not sufficient to take into account only the complexity of the algorithm, i.e., the number of floating point operations, but also the overhead stemming from the underlying data structures. Two main problems arising in the implementation of adaptive schemes are (i) *dynamic memory operations* and (ii) *fast data access* with respect to inserting, deleting and finding an element. In the following these two issues shall be briefly discussed.

Dynamic memory operations. Due to refinement and coarsening operations in the algorithm, memory space for new elements or freeing unused elements are frequently performed and therefore should be very fast. Since the memory operations provided by the programming language are universal, they can be made much more efficient if we take into account that our data are of the same type. In particular, this can be realized by allocating a sufficiently large memory block and manage the memory requirements by the algorithm with a specific data structure. Since the overall memory demand can only be estimated, the data structure should provide dynamic extension of the memory.

Fast data access. For instance, lists might be considered as a container for fast inserting and deleting elements. However, finding an element requires the search in the whole list from its beginning until the position of the element is found. On the other hand, a fast way of finding an element by means of an index is an array. But inserting and deleting elements in an array requires copy operations. However, if one does not need to maintain any *ordering in the data*, then other data structures, e.g. hash maps, might be a good alternative.

In view of an optimal memory management and a fast data access we design a *hash map* that is composed of two parts, namely, a vector of pointers, a so-called *hash table*, and a memory heap, see Fig. 6.1. The hash table is connected to a *hash function* $f : I \to \mathcal{P}$ which maps a *key*, e.g. an index, to a position on the hash table, i.e., a number between 0 and $\#\mathcal{P} - 1$ where $\#\mathcal{P}$ denotes the length of the hash table. Note, that in general the set of all possible keys is much larger than the length of the hash table, i.e., $\#\mathcal{P} \ll \#I$. Obviously, the hash function is not injective. This leads to collisions in the hash table, i.e., different keys might be mapped onto the same position by the hash function. The corresponding values of these keys are then linked to the list that starts at the position $f(key)$. Each element in the hash table is a pointer to a connected list whose elements are stored in the heap. Here each element of the list can be a complex data structure itself. It contains at least the key and usually additional data, the so–called *value*. In general, the value consists of the data corresponding to an index. In addition, the level might be additionally provided. In order to maintain the connection of the value to its key, the key also has to be stored in this list.

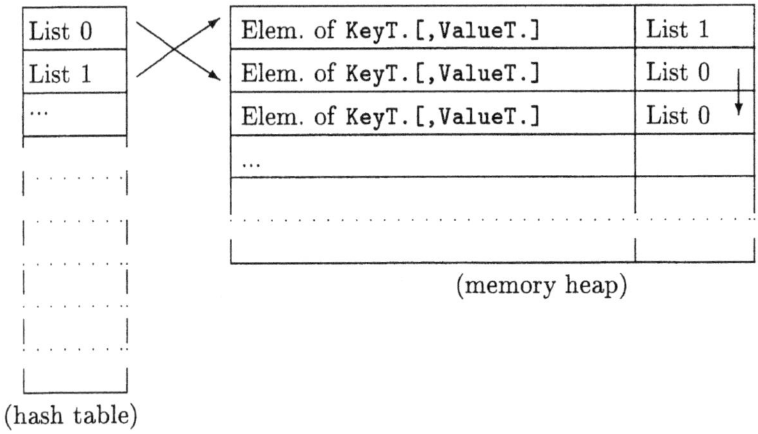

Fig. 6.1. Hash map

Obviously, the performance of the hash map crucially depends on the number of collisions, since the access to the values by the linked list is the faster the shorter the list. To this end, the number of keys mapped to the same position in the hash table should be as small as possible. This is controlled by the *fill rate* of the hash table, i.e., the probability of collisions. If $\#\mathcal{P} = \#I$ and f is injective, then no collisions occur. However, this case leads to a tremendous waste of memory space in the hash table, since the adaptive scheme will only access a small number of keys $\mathcal{I} \subset I$ with $\#\mathcal{I} \ll \#I$. On the other hand, if the length of the hash table is too small this might cause a lot of collisions. In order to balance the length of the hash table and the number of collisions, several strategies have been developed for the design of an appropriate hash function, see [Str97, OW96]. For our purpose choosing the modulo function as the hash function turned out to be sufficient. It maps an integer generated from a key to a position $f(i) \in \mathcal{P} = \{0, \ldots, p-1\}$ in the hash table by

$$f(i) = i \bmod p.$$

As already discussed the length p of the hash table crucially influences the performance of the hash map. In general, p is chosen to be the largest prime number less than $\#\mathcal{I}$ in order to reduce the probability for a collision. However, in our computations it turned out that the choice of p as a prime number is not significant. Therefore we choose a simpler strategy. To this end, we guess the number of expected elements to be stored, say `nMax`, and choose the length of the hash table as `nMax * nFactor`, where `nFactor` equals 3 in our computations. At the same time we allocate memory for only `nMax` elements in the associated heap. This implies that we work with a hash table filled about `dFill = 1 / nFactor` but allocate an optimal chunk of memory for our data. This strategy combines the advantages of a large hash table, i.e., small number of collisions, with the requirement to allocate only adequate storage memory.

So far we have outlined the idea of hashing as it can be found in standard text books. However, for our applications we have to modify the standard concept, since we need to know the connection of the values to a certain level $j \in \{0, \ldots, L-1\}$. Remember that the local transformations are performed level by level. For this purpose, the hash map is extended by a vector where its length is determined by the number of levels L. The jth component of this vector contains a pointer that points to the first element of level j put into the memory heap. Additionally, the value has to be internally extended by a pointer that points onto the next element of level j. This is sketched in Fig. 6.2. Then we can access all elements of level j by traversing the resulting connected list.

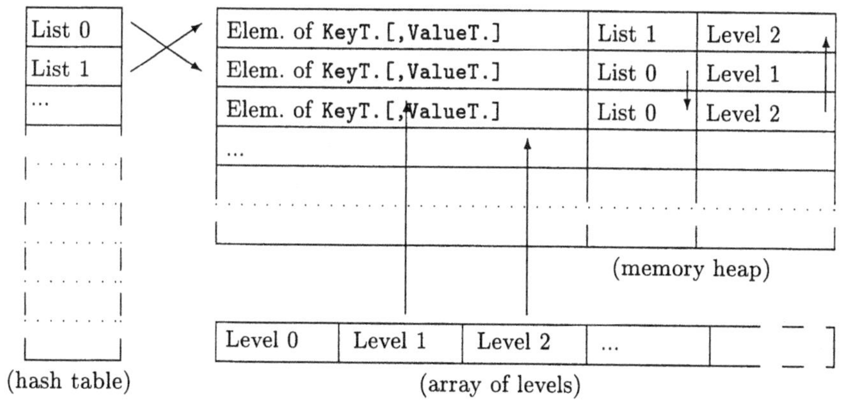

Fig. 6.2. Linked hash map

6.3 Data Structures

In the following we describe the data types that are needed for the implementation of the Algorithms 3 and 6 of the local transformations. Here we restrict ourselves to the curvilinear case as summarized in the Sect. 3.8. The data types are derived from template classes provided by the template library igpm_t_lib, see [MV00].

First of all we derive the array types

```
typedef tvector_n< double,  m    >   uvector
typedef tvector_n< double,  2^d-1>   evector
typedef tvector_n< uvector, 2^d-1>   dvector
```

from the template class tvector_n by which we can store the vector of conservative quantities $\mathbf{v}_{j,\mathbf{k}} \in \mathsf{R}^m$, the details $\mathbf{d}_{j,\mathbf{k},\mathbf{e}} \in \mathsf{R}^m$ and the vector of details corresponding to all wavelet types $(\mathbf{d}_{j,\mathbf{k},\mathbf{e}})_{\mathbf{e}\in E^*}$.

In the curvilinear case, the averages and the details are enumerated by multiindices $\mathbf{k} = (k_1, \ldots, k_d) \in \mathsf{N}_0^d$ and level–multiindices $(j, \mathbf{k})$. To this end, we derive the integer arrays

```
typedef tmultiindex< d >          mi
typedef tlevelmultiindex< mi >   lmi
```

from the template classes tmultiindex and tlevelmultiindex, respectively.

For the local transformations we always access simultaneously all non–vanishing elements of a column of the mask matrices. Therefore it is convenient to agglomerate the indices of the corresponding supports in one data structure. However, we do not have to store index by index, since in the curvilinear case the supports can be represented as a multidimensional integer interval $\mathbf{k}+[0,i]^d \subset \mathsf{N}_0^d$ characterized by the multiindex $\mathbf{k}$ and the number i. By this knowledge we designed the special template class tmultirange. According to the supports of the box function and the modified box wavelets, see Subsection 3.8, we need four different types of multiranges

```
typedef tmultirange< 1,      mi>   mr1
typedef tmultirange< 2,      mi>   mr2
typedef tmultirange< 4*s+2, mi>   mr4s
typedef tmultirange< 2*s+1, mi>   mr2s
```

for the supports $\check{\mathcal{M}}^{*,0}_{j,\mathbf{k}}$ $(\check{\mathcal{G}}^0_{j,\mathbf{k}})$, $\check{\mathcal{M}}^0_{j,\mathbf{k}}$ $(\check{\mathcal{G}}^{*,0}_{j,\mathbf{k}})$, $\mathcal{M}^e_{j,\mathbf{k}}$ and $\mathcal{L}^s_{j,\mathbf{k}}$, respectively. Here the first argument specifies the length i and the second argument the initial index $\mathbf{k}$.

Analogously, we can store the non–vanishing matrix elements corresponding to the columns of the mask matrices by multiranges where in addition to the blocked multiindices the corresponding matrix elements are stored whose type is specified by the third argument. Here we distinguish between the types

```
typedef tmultirange< 1, mi, double >mrA_G0c
typedef tmultirange< 2, mi, double >mrA_M0c
```

for the matrix columns of $\check{\mathsf{G}}_{j,0}$ and $\check{\mathsf{M}}_{j,0}$, respectively. In case of the matrices $\mathsf{L}_{j,\mathbf{e}}$, $\mathbf{e} \in E^*$, we proceed differently. By construction the supports $\mathcal{L}^{\mathbf{e}}_{j,\mathbf{k}}$ are the same for all wavelet types $\mathbf{e} \in E^*$, only the matrix elements $l^{j,\mathbf{e}}_{\mathbf{r},\mathbf{k}}$ depend on the wavelet type. Moreover, in the local transformations we always have to access all types corresponding to an index pair $(j, \mathbf{k})$. Therefore it is more convenient to store all matrix elements $l^{j,\mathbf{e}}_{\mathbf{r},\mathbf{k}}$, $\mathbf{e} \in E^*$, in one multirange

```
typedef tmultirange< 2*s+1, mi, evector >     mrA_L .
```

In this we avoid the multiple storing of the index $\mathbf{k}$ and the length i of the integer interval.

The mask matrices $\check{\mathsf{M}}_{j,0}$, $\check{\mathsf{G}}_{j,0}$ and $\mathsf{L}_j = (\mathsf{L}_{j,\mathbf{e}})_{\mathbf{e}\in E^*}$ are then stored in hash maps, see Sect. 6.2, where each element is a multirange representing the non–vanishing matrix elements corresponding to one matrix column

```
typedef thashmap_linked< lmi, mrA_M0c >  liMap_M0c
typedef thashmap_linked< lmi, mrA_G0c >  liMap_G0c
typedef thashmap_linked< lmi, mrA_L   >  liMap_L   .
```

Here the first argument of the hash map represents the type of the key, e.g. a level–multiindex, and the second argument the type of the value, e.g. a multirange.

Analogously the local averages $\mathbf{v}_{j,\mathbf{k}}$, $(j, \mathbf{k}) \in \mathcal{G}_{L,\varepsilon}$, and the significant details $\mathbf{d}_{j,\mathbf{k},\mathbf{e}}$, $(j, \mathbf{k}, \mathbf{e}) \in \mathcal{D}_{L,\varepsilon}$, are stored in linked hash maps

```
typedef thashmap_linked< lmi, uvector >    liMap_u
typedef thashmap_linked< lmi, dvector >    liMap_d ,
```

where each element is either a vector of conservative quantities or an array of vectors representing the details of all conservative quantities and all wavelet types. Note, that the index sets $\mathcal{I}_{j,\varepsilon}$ and $\mathcal{J}_{j,\varepsilon}$ are implicitly determined by the linked lists corresponding to the different levels.

The local transformations are performed levelwise. To this end, we need data structures where temporary data corresponding to one level can be stored. Again we use hash maps. Since only *one* level is involved, we can simplify this type of hash map. Therefore the data types

```
typedef thashmap_linked_one<mi          >   iSet
typedef thashmap_linked_one<mi, uvector >   iMap_u
typedef thashmap_linked_one<mi, dvector >   iMap_w
```

use the template class `thashmap_linked_one`.

Note, that the hash map has to be linked, since we have to traverse through all elements. In particular, the data type `iSet` is a hash map where only keys are stored but no values. This can be interpreted as a set of keys, e.g. multiindices.

Finally we need a data structure for the management of the adaptive grid and corresponding geometric information. For instance, to each cell volume, monomials of higher order, normals, arc lengths etc. have to be stored. Which information are needed does not only depend on the multiscale setting but also on the FVS. We therefore do not specify the details but collect the geometric data in a class of its own

```
class cell_data { ... }.
```

Then the adaptive grid is stored in the hash map type

```
typedef thashmap_linked<lmi,cell_data> liMap_grid
```

where each element is an element of the class `cell_data`.

By means of the above data structures we now introduce the hash maps

```
liMap_M0c   M0c
liMap_G0c   G0c
liMap_L     L
```

for the management of the mask matrices $\check{\mathsf{M}}_{j,0}$, $\check{\mathsf{G}}_{j,0}$ and L_j, $j = 0, \ldots, L-1$, as well as the local averages and the significant details

```
liMap_u   v_map
liMap_d   d_map
```

corresponding to the adaptive grid and the set of significant details. The adaptive grid and related geometric information are stored in the hash map

```
liMap_grid   adapgrid_map
```

where not only the locally finest cells are taken into account but also all cells on coarser levels.

The initialization of these hash maps crucially influences the performance of the computation. We therefore describe the choice of `nMax`, see Subsection 6.2, which determines the length of the hash table and the size of the memory needed for storing `nMax` elements. First of all, we consider the mask matrices. We observe that the number of elements corresponds to the number of columns of these matrices, since each element in the hash map represents the non–vanishing matrix coefficients corresponding to one column. For the matrices $\check{\mathsf{M}}_{j,0}$ and L_j the total number of columns for all levels is determined by

$$\sum_{j=0}^{L-1} N_j = N_{L-1} \, \frac{1-q^L}{1-q} \quad \text{with} \quad q = (\# E)^{-1},$$

since $N_j = q\, N_{j+1}$. In case of the matrix $\check{\mathsf{G}}_{j,0}$ we obtain the total number

$$\sum_{j=0}^{L} N_j = N_L \, \frac{1-q^{L+1}}{1-q}.$$

This is also an upper bound for the hash map `adapgrid`, since we store not only cell information corresponding to the adaptive grid but also the information for all coarser cells. For the averages and the details the total number of elements is restricted to N_L for the full grid on the finest level and accordingly all details are significant, i.e., $N_{L-1}\,(1-q^L)/(1-q)$.

From the total number of elements that have to be stored in the worst case of a uniform refinement over all levels, we determine the number `nMax = nTotal * rFill` of predicted elements where `rFill` denotes the fill rate. Note, that `rFill` differs from the fill rate `dFill` of the hash table. In Sect. 7.1.4 we will investigate the influence of `rFill` and `nFactor` on the performance of the computation.

Finally we would like to remark that the hash maps for the mask matrices as well as the adaptive grid are initialized once and then the length of the hash table as well as the memory heap size remain unchanged throughout the computation. Therefore the choice of `rFill` is more significant for the mask matrices and the adaptive grid. In particular, the corresponding hash maps require the bulk of memory. This is different for the hash maps in which the local averages and the details are stored, since the adaptive grid and the set of significant details, respectively, are changing from one time step to the next one. Here the length of the hash table has to be adapted in each time step and the memory eventually has to be extended dynamically. From Algorithm 2 we conclude that the complexity of the sets $\mathcal{G}_{L,\varepsilon}$ and $\mathcal{D}_{L,\varepsilon}$ are related by $\# \mathcal{G}_{L,\varepsilon} = q^{-1} \# \mathcal{D}_{L,\varepsilon}$, since a cell is refined as long as there exists a significant detail. Then we can adapt the size of the hash maps `v_map` and `d_map` to the actual number of elements determined by one of the hash maps with `nMax` = $q^{-1} \# \mathcal{D}_{L,\varepsilon}$ for `v_map` and `nMax` = $q \# \mathcal{G}_{L,\varepsilon}$ for `d_map`, respectively.

7 Numerical Experiments

In the previous chapters we have developed and investigated a new concept for the construction of an adaptive FVS where we apply local multiscale techniques to a reference scheme. For scalar conservation laws we have been able to derive an error estimate of the form (5.1). This is based on an a priori error estimate for the *discretization error* of the reference FVS and the *stability of the perturbation error* in the sense of (5.3). In [CKMP01] parameter studies have been presented for scalar one–dimensional problems which confirm the analytical results in Chapter 4. In the sequel, we verify that the adaptive concept can also be applied to *systems of conservation laws*. In particular, we are interested in applications to *real–world problems* arising from problems in engineering. For this purpose, we present several computations for the two–dimensional Euler equations for a polytropic gas with $\gamma = 1.4$, see Example 3. These have been carried out on PC's with a 600 MHz processor (Pentium III).

The numerical investigations are divided into two parts. In the first part we are concerned with the computational complexity of the adaptive scheme with respect to computational costs and memory requirements as well as the stability of the perturbation error. To this end, we present several parameter studies for three test configurations, namely, (i) shock reflection, (ii) implosion and (iii) wave interaction. These are instationary problems performed on Cartesian grids. In the second part we report on some computations for real–world applications. Here we consider (i) the flow over a bump and (ii) the flow around a profile of an airfoil. Both configurations result in steady state solutions. Since the geometry of the computational domains is no longer that simple as before, we discretize the flow fields by block–structured curvilinear grids.

7.1 Parameter Studies

The analytical setting of Theorem 5 is much too restrictive with regard to realistic problems. Therefore we are interested in the performance of the adaptive scheme applied to an extended setting not necessarily covered by the analytical investigations. Here three issues are of special interest, namely, (i) the detection of all kinds of singularities, e.g., discontinuities and kinks, corresponding to physically relevant effects by means of the multiscale decomposition, (ii) the complexity of the adaptive FVS with respect to computational costs and memory requirements in comparison to the reference FVS and (iii) the stability of the perturbation error. These issues are investigated by means of several parameter studies with respect to an increasing number of refinement levels where the threshold value ε is fixed for all computations. Note, that this does not quite agree with the ideal strategy as outlined in the beginning of Chapter 5. In order to balance the discretization error and the perturbation error, the threshold value ε has to depend on the number of refinement levels L according to (5.1) and (5.2). However, for systems of conservation laws no error estimate is available so far for the discretization error. On the other hand, the stability of the perturbation error in the sense of (5.3) has only been verified for scalar problems. Therefore in the following we focus in our numerical investigations on the stability of the perturbation error, in particular, with regard to its dependence on the highest resolution level L.

First of all, we summarize the setting of the test configurations. Then we specify the discretization and present the reference FVS. By means of parameter studies we investigate and discuss the computational complexity of the adaptive FVS as well as the stability of the perturbation error for an increasing number of refinement levels L. We conclude with some parameter studies concerning the parameters `rFill` and `nFactor` of the underlying data structures.

7.1.1 Test Configurations

We consider three configurations where the computational complexity of the problem strongly varies due to the singularities occurring in the solution.

Shock Reflection. This is a quasi–one-dimensional problem where we impose a single shock aligned with one of the coordinate axes, see Figure 7.1. The shock is moving to the right approaching a wall. At the wall the shock is reflected and moves back to the left. Note, that the initial data in front of

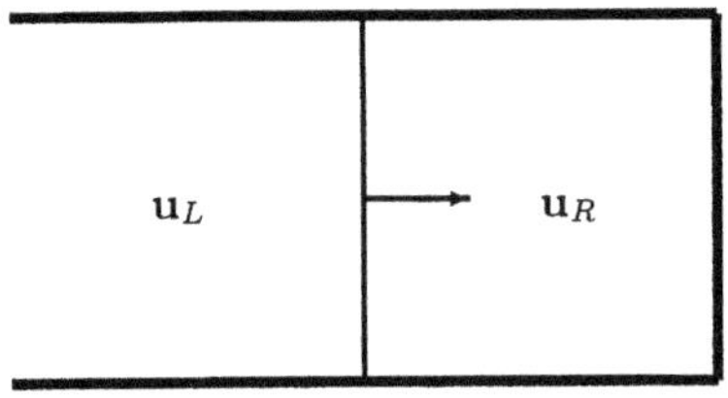

	Units	State L	State R
ϱ	kg/m^3	1.3	0.8
u_x	m/s	1300	0
u_y	m/s	0	0
p	hPa	11050	5200

Fig. 7.1. Initial configuration for shock reflection

the shock correspond to a supersonic state.

Implosion Problem. The initial configuration is determined by two states $\mathbf{u}_E$ and $\mathbf{u}_I$ as well as the radius r of a circle, see Figure 7.2. Inside the circular domain we impose low pressure and outside high pressure. Again the resulting flow field is quasi–onedimensional due to the rotational symmetry. With propagating time three waves develop, namely, (i) a rarefaction wave, (ii) a contact surface and (iii) a shock wave. The shock wave and the contact surface are moving towards the center of the circle. The rarefaction wave is moving into the opposite direction where the corresponding rarefaction fan is expanding. Here we are interested in the instance when the shock wave focuses in the center.

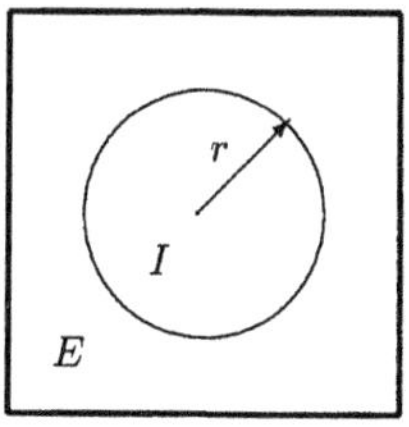

	Units	State I	State E
ϱ	kg/m^3	1.251	2.502
u	m/s	0	0
p	Pa	101280	202560
r	m	0.15	

Fig. 7.2. Initial configuration for implosion problem

Wave Interaction. The initial configuration is determined by four states corresponding to the four quadrants of the coordinate system, see Figure 7.3. Away from the origin of the coordinate system, the solution exhibits a one–dimensional wave pattern consisting of a rarefaction wave, a contact surface and a shock wave. This structure is obtained by the solution of a one–dimensional Riemann problem characterized by the constant states $\mathbf{u}_L = \mathbf{u}_I$, $\mathbf{u}_R = \mathbf{u}_{II}$ and $\mathbf{u}_L = \mathbf{u}_{II}$, $\mathbf{u}_R = \mathbf{u}_I$, respectively. Close to the origin the different one–dimensional waves interact forming a genuinely two–dimensional wave pattern.

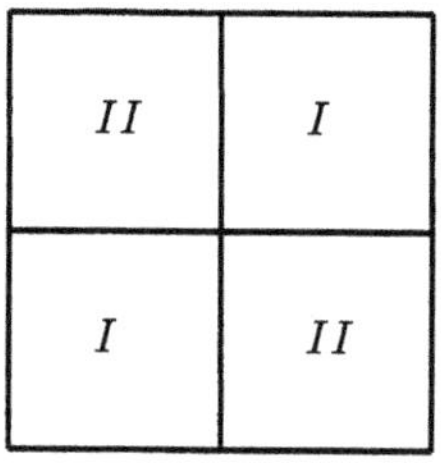

	State I	State II
ϱ	0.125	1.000
u_x	1.000	0.000
u_y	1.000	0.000
p	0.100	1.000

Fig. 7.3. Initial configuration for wave interaction

7.1.2 Discretization

As the reference FVS we apply an essentially non–oscillatory (ENO) scheme described in [Mül93]. This ENO scheme is characterized by a one–dimensional second order accurate reconstruction technique via primitive functions, see [HEOC87]. Here the primitive variables are reconstructed by means of piecewise linear polynomials. At the cell interfaces one–dimensional Riemann problems in normal direction are solved applying Roe's approximate Riemann solver, [Roe81].

For all three configurations the boundary conditions are simple. In principle, three types of boundary conditions occur, namely, (i) *free stream conditions* at inflow boundaries, see the left side boundary in Figure 7.1, (ii) *slip conditions* at a wall, see the right side boundary in Figure 7.1 and (iii) *attached flow field* whenever the flow field at the boundary is constant in normal direction.

The discretization of the computational domain is summarized in Table 7.1. Since we employ an explicit FVS, the step size τ in time has to satisfy a CFL condition for the *finest* discretization level L, i.e.,

$$\lambda_{L,k}\, \max\{|\lambda_k(\mathbf{u},\mathbf{n})| \; ; \; \mathbf{u}\in\mathcal{D},\ \|\mathbf{n}\|_2 = 1\} \le c < 1.$$

Note, that for a uniform discretization $|V_{j,k}| = \prod_{i=1}^{d} h_{j,i} = h_j$ and $h_{j,i} = 2^{-1}h_{j-1,i} = 2^{-j}h_{0,i}$. For the parameter studies we successively increase the number of refinement levels L. In order to maintain the CFL number we have to reduce τ by the factor 0.5 when adding an additional refinement level, i.e., $\tau \equiv \tau_L = 2^{-1}\tau_{L-1} = \ldots = 2^{-L}\tau_0$. From this we conclude $\lambda_{L,k} = 2^{(d-1)L}\tau_0/h_0$. Obviously, the step size τ is characterized by the number of refinement levels L and the ratio τ_0/h_0 of the coarsest discretization listed in Table 7.1.

Table 7.1. Discretization parameters

Case	Ω [m]	T [s]	τ_0/h_0 [s/m^2]
Reflection	$[0,1]\times[0,0.4]$	3.2×10^{-4}	1.28×10^{-4}
Implosion	$[0,1]^2$	3.6125×10^{-4}	10^{-6}
Interaction	$[-0.4,0.4]^2$	1.5×10^{-1}	2.5

The multiscale analysis is based on the modified box wavelets. These are constructed according to Algorithm 1 where the degree of vanishing moments is chosen as $M = 3$. In the curvilinear case, the stencils $\mathcal{L}^e_{j,k}$ are determined by $\mathcal{L}^s_{j,\mathbf{k}}$ with $s = 1$ according to (3.45) and (3.46). The sequence ε of threshold values is recursively determined by $\varepsilon_L = \varepsilon$, $\varepsilon_j = 0.5\,\varepsilon_{j+1}$, $j = L-1,\ldots,0$, according to Proposition 3. Here the threshold value is $\varepsilon = 0.001$ (implosion, wave interaction) and $\varepsilon = 0.01$ (shock reflection) for all computations of the parameter study.

Finally, we have to specify the flux evaluation. Here we use the locally structured strategy (4.15) because the reference numerical flux is only defined on a *structured* grid. For this purpose, the adaptive grid has to be locally uniform of degree $p = 3$ according to the flux stencil. From Corollary 5 we conclude that this is guaranteed if the tree of significant details is graded of degree $q = 2$.

7.1.3 Computational Complexity and Stability

For the three test configurations we have performed parameter studies with respect to an increasing number of refinement levels where the threshold value ε is fixed for all computations. In Figures A.1, A.2, A.5, A.6, A.9, A.10 the density is presented for the initial data $t = 0$ and the time $t = T$ listed in Table 7.1. The adaptive grids are plotted in Figures A.3, A.4, A.7, A.8, A.11, A.12. All these figures reflect the computational results for the highest refinement levels listed in the Tables 7.2, 7.3, 7.4. We observe that the higher resolution levels are only accessed near discontinuities. This verifies that the adaptation criterion based on details is able to detect the relevant physical effects in the flow field.

We are now interested in the computational complexity of our scheme (with regard to computational time and memory) for the present studies of stability. This is documented in the Tables 7.2, 7.3, 7.4. The quantities listed in the different columns are

- the number of refinement levels L,
- the number N_L of all cells corresponding to the full grid on level L,
- the number $N_{\mathcal{G}} := \#\mathcal{G}_{L,\varepsilon}/N_L$ representing the number of cells corresponding to the largest adaptive grid determined during the computation relative to the full grid on level L,
- the computational time C_{MS} used by the adaptive scheme,
- the memory size Mem which corresponds to the maximum amount of memory that has been allocated in one time step,
- the speedup rates $S_{MS} := C_{FVS}/C_{MS}$ determined by the ratio of the computational times for the adaptive scheme and the reference FVS on the full grid of level L and
- the perturbation error $\mathbf{e}_L := \overline{\mathbf{v}}_L - \mathbf{v}_L \in \mathsf{R}^{N_L}$ determined by the difference of the averages $\overline{\mathbf{v}}_L := \{\overline{\mathbf{v}}_{L,k}\}_{k\in I_L}$ and $\mathbf{v}_L := \{\mathbf{v}_{L,k}\}_{k\in I_L}$ produced by the adaptive scheme and the reference FVS on the full grid of level L, respectively. The local averages from the adaptive scheme are mapped onto the full grid by means of local reconstruction according to the inverse multiscale transformation.

Table 7.2. Parameter study for shock reflection

L	N_L	$N_{\mathcal{G}}$ %	C_{MS} [min]	Mem [MB]	S_{MS}	$\|\|\mathbf{e}_L\|\|_{1,L,*}$
3	2560	31	1.50E−1	0.9	2.1	1E−3
4	10240	16	6.84E−1	1.8	3.7	6E−4
5	40960	8	3.03E+0	3.6	6.8	4E−4
6	163840	4	1.36E+1	7.3	12.6	2E−4
7	655360	2	5.98E+1	14.5	21.8	1E−4
8	2621440	1	3.04E+2	29.0	34.3	

Table 7.3. Parameter study for implosion

L	N_L	$N_{\mathcal{G}}$ %	C_{MS} [min]	Mem [MB]	S_{MS}	$\|\|\mathbf{e}_L\|\|_{1,L,*}$
4	16384	45	6.833E−1	8	1.34	1E−3
5	65536	28	3.667E+0	20	2.10	6E−4
6	262144	14	1.603E+1	40	3.61	4E−4
7	1048576	7	6.112E+1	84	7.73	4E−4
8	4194304	4	2.194E+2	165	16.51	
9	16777216	2	7.470E+2	334	36.44	

Table 7.4. Parameter study for wave interaction

L	N_L	$N_{\mathcal{G}}$ %	C_{MS} [min]	Mem [MB]	S_{MS}	$\|\|\mathbf{e}_L\|\|_{1,L,*}$
2	6400	89	0.73E+0	6	0.82	1.2E−3
3	25600	58	4.53E+0	16	0.98	1.1E−3
4	102400	34	2.33E+1	39	1.62	1.2E−3
5	409600	19	1.12E+2	90	2.67	1.2E−3
6	1638400	10	4.99E+2	186	4.79	
7	6553600	5	1.98E+3	337	9.67	

Computational complexity. Considering the three tables we make the following observations:

- The number N_L increases by a factor of 2^d because the grid is uniformly refined by subdividing a cell into 2^d congruent subcells.
- The relative number $N_{\mathcal{G}}$ decreases asymptotically by a factor $r_L := N_{\mathcal{G},L}/N_{\mathcal{G},L-1}$. Therefore the total number of cells in the adaptive grid increases by a factor $q_L := \#\mathcal{G}_{L,\varepsilon}/\#\mathcal{G}_{L-1,\varepsilon} = 2^d r_L$. In general, the ratio

r_L is smaller than 0.5 and, hence, the number of cells in the adaptive grid increases only by a factor of 2 at most with each additional refinement level instead of 2^d for the uniform refinement. This results in an exponential reduction of cells in comparison to the full grid that corresponds to significant speedup rates. However, we would like to emphasize that the full grid may give us the more accurate approximation.

- The computational time C_{MS} increases with $L \to \infty$ because the number of time steps is doubled according to the CFL condition for the finest level. On the other hand, the number of cells in the adaptive grid increases by the factor q_L. Therefore C_{MS} increases by the factor $2\,q_L$ which is smaller than the 2^{d+1} for the reference FVS on the finest level. This results in an exponential increase of the speedup rates S_{MS}, i.e., the computational time becomes significantly smaller for the adaptive scheme in comparison with the computation of the reference FVS on the full grid.
- Since the number of cells in the adaptive grid increases by the factor q_L, we have to allocate more memory space. This is reflected by the quantity Mem listed in the tables. Comparing these numbers we conclude that memory size is increasing by a factor proportional to q_L. We emphasize that for Cartesian grids the mask coefficients do not depend on the level and the position except for boundary adaptations and, therefore, require much less memory. Additionally, these values can be computed in a preprocessing step. This is not taken into account here.

In principle, these observations hold for all the three test configurations although the numbers differ according to the wave structures developing in the different flow fields. Here the shock reflection gives the highest speedup rates whereas the wave interaction exhibits the lowest rates due to the complex wave structures. Therefore we conclude that the computational effort and the memory requirements are proportional to the number of cells in the adaptive grid. We emphasize that this can *not* be realized by Harten's original strategy because the resulting scheme still involves the full grid on the finest level. This is based on the fact that locally expensive numerical fluxes are replaced by cheaper approximations, i.e., Harten's strategy is based on a *hybrid* flux evaluation where the switching is performed by means of the multiscale analysis. However, the adaptive strategy at hand reduces the *total* number of flux evaluations not only the number of *expensive* flux evaluations, i.e., we do not compute the cheap flux approximations in smooth regions at all. In combination with the *local* multiscale transformation and its inverse, the number of floating point operations as well as the memory size of the resulting scheme is proportional to the number of cells in the adaptive grid.

Stability. The objective of the adaptive strategy is to reduce the computational complexity while maintaining the accuracy in comparison to the reference scheme on a uniform grid. In order to balance the discretization error and the perturbation error in (5.1) according to Corollary 7 we need the stability of the perturbation error in the sense of (5.3). We would like to emphasize that the analytical results have only been verified for one–dimensional scalar problems, see Section 5.3. Here we consider a two–dimensional system. In addition, the numerical flux of the reference FVS does not match the assumptions of Theorem 5. Since for systems of conservation laws bounds on the global discretization error are not available we focus in the following just on the stability of the compression scheme. That is, instead of balancing discretization and perturbation error we fix the threshold value ε and explore the dependence of the perturbation error on the level L, the number of time steps n and the temporal step size τ. For this purpose, the perturbation error is also listed in the Tables 7.2, 7.3, 7.4 as far as the computation could be performed on the available computers. Here the error $\mathbf{e}_L$ is measured in the weighted l^1–metric where each conservative quantity is scaled by its global maximum in the computational domain, i.e.,

$$\|\mathbf{e}_L\|_{1,L,*} := \max_{i=1,\ldots,m} \|\mathbf{e}_{L,i}\|_{1,L}/\|\mathbf{e}_{L,i}\|_{l^\infty}$$

with $\mathbf{e}_{L,i} := (e_{L,k,i})_{k \in I_L}$. Recall that the thresholding is also performed by means of the scaled details, see (3.38). We observe that the error is proportional to the threshold value ε. For the shock reflection this error is even significantly smaller because the solution is piecewise *constant*. Hence no error is introduced by the truncation in regions where the solution is constant. To some extent, this also holds for the implosion. In case of the wave interaction the solution is no longer piecewise constant but only piecewise smooth exhibiting a genuine two–dimensional structure.

We conclude that for all test cases the perturbation error does not increase with increasing number of refinement levels L. Note, that for an additional refinement level the number of time steps is doubled and the temporal step size is halved according to the CFL constraint. An additional error is introduced because the local flux evaluation is performed by (4.15) and not by (4.13) which is used in the analysis. Investigations in [CKMP01] show that this does not affect the perturbation error provided the reference FVS is higher order accurate. This no longer holds for a first order FVS. In this case a significant error is observed.

We conclude the discussion on the parameter studies with a remark on the "ideal" computation in the sense of (5.2). Whenever the perturbation error is smaller than the discretization error, we waste performance in the sense of computational time and memory size, because the discretization error is still dominating. Conversely, we loose accuracy if the perturbation error is larger than the discretization error. Therefore the ideal computation of the parameter study corresponds to the refinement level where both the perturbation error and the discretization error are balanced.

Finally, we would like to discuss the limitations of the adaptive scheme. For this purpose, we present some plots of the error, see Figures A.13, A.14 and A.15. We observe that the error accumulates near discontinuities. This seems to be a contradiction because near discontinuities the adaptive scheme is expected to locally refine the grid. However, errors introduced away from the discontinuities are transported to a shock layer by means of the characteristics intersecting there. On the other hand, the reference FVS is known to smear contact discontinuities. This leads to a smoothing of the solution which is detected by the multiscale analysis, i.e., near contact discontinuities the adaptive scheme does not refine the grid until the *finest* level is reached. This confirms that the adaptive scheme can, of course, not improve the solution of the reference FVS but it can only provide a perturbation of this solution in a much more efficient way.

The previous discussion indicates that the adaptive scheme meets the design criteria, namely, (i) the computational effort and the memory requirements are proportional to the number of cells in the adaptive grid and (ii) the perturbation error is proportional to the threshold value. Since in our parameter studies the threshold value ε is fixed, i.e., $\varepsilon \neq \varepsilon(L)$, we can not directly conclude on the discretization error. If we *assume* that the discretization error for the uniform grids of level L behaves like $2^{-\alpha L}$ we see from the tables at which level the recorded perturbation error is of comparable size which gives an impression of the interrelation between the overall accuracy and the amount of computational work. However, the computations on the full uniform grid of level L can only be realized for a small number of refinement levels L due to the huge amount of memory needed.

7.1.4 Hash Parameters

The efficiency of the computations is significantly influenced by the initialization of the hash maps `M0c`, `G0c`, `L` and `v_map`, `d_map` as well as `adapgrid_map` by which we store the mask matrices, the sequences of local averages and significant details as well as the adaptive grid. For this purpose, we carry out two parameter studies concerning the parameters `rFill` and `nFactor`, see Section 6.3, for the implosion configuration. Here one of the two parameters is fixed while the other varies, see Table 7.5 and 7.6. We are primarily interested in the impact of these parameters on the computational time, the size of allocated memory and the actual fill rates of the above hash maps.

First of all, we investigate the influence of the parameter `nFactor` by which the length of the hash table is varied. The larger `nFactor` is chosen the smaller becomes the fill rate `dFill` of the hash table, i.e., the probability for a collision decreases, and visa versa. For this parameter study `rFill` is fixed by 1, i.e., the size of the allocated memory heap for storing the keys and values is that of the worst case corresponding to the uniform finest grid. Considering Table 7.5 we note that the memory size increases with `nFactor` because the hash table becomes larger. At the same time, the actual fill ratio of the hash maps is reduced. However, choosing `nFactor` larger than 3 does not significantly affect the computational time. Conversely, if we reduce `nFactor` then the size of the hash table becomes smaller which results in

Table 7.5. Implosion: parameter study for **nFactor** with **rFill** = 1

nFactor	1	2	3	4
$L = 4$				
CPU [sec]	51	50	49	50
fill rate [%]	58	29	19	15
Memory [MB]	7.9	8.5	9.2	9.2
$L = 5$				
CPU [sec]	249	238	233	235
fill rate [%]	40	20	13	10
Memory [MB]	23	25	26	31
$L = 6$				
CPU [sec]	1068	1044	1030	1032
fill rate [%]	29	15	10	7
Memory [MB]	64	71	75	79

more collisions. This leads to a significant increase of the computational time because accessing data in the hash maps consumes more time due to collision management. The results in Table 7.5 suggest to choose `nFactor` = 3 in view of an optimal computational time.

Table 7.6. Implosion: parameter study for **rFill** with **nFactor** = 3

$L = 4$					
rFill	0.12	0.25	0.5	1	2
CPU [sec]	50	49	49	49	49
fill rate [%]	161	77	39	19	10
Memory [MB]	7.6	8.0	7.9	9.3	12
$L = 5$					
rFill	0.08	0.17	0.35	0.7	1.4
CPU [sec]	238	234	236	234	234
fill rate [%]	166	78	38	19	10
Memory [MB]	19	18	19	22	31
$L = 6$					
rFill	0.04	0.08	0.15	0.3	0.6
CPU [sec]	1060	1035	1029	1036	1026
fill rate [%]	243	121	65	33	16
Memory [MB]	36	38	38	40	55

Another parameter study was performed for `rFill` predicting the relative size of the adaptive grid in comparison to N_L of the uniform finest grid. This factor influences the size of the memory heap for storing the keys and values as well as the length of the hash table. Considering Table 7.6 we note that the size of allocated memory is the larger the higher the value for `rFill`. At the same time, the fill rate of the hash tables decreases, i.e., the number of collisions reduces. This has a positive effect on the computational time. The actual fill rates for the implosion problem can be deduced from Table 7.3 as 0.45, 0.28 and 0.14 for the refinement levels $L = 4$, 5 and 6. Then we conclude from Table 7.6 that the memory size and the computational time are balanced if we choose these numbers for `rFill`. Since in general these numbers are not available before the computation has been carried out, the value `rFill` has to be predicted. Here the parameter study suggests that the predicted value should be at least as high as the real value, i.e., overpredicting `rFill` is preferable to underpredicting. If it is smaller then the computational time increases significantly due to a higher collision rate.

Finally, we would like to remark that the collision rate can be reduced if an optimal prime number, i.e., a good choice for a modulo number, is chosen for the length of the hash table. However, investigations in the context of the adaptive scheme show that the effort for determining an optimal number is much more expensive than collisions caused by our choice for the modulo number.

7.2 Real World Application

Characteristic for real world applications are *multidimensional* configurations with *complex* geometries. These configurations can only be adequately discretized by means of unstructured or block–structured grids, respectively. Although unstructured grids, e.g. arbitrary triangulations, are easier to generate we prefer structured grids, because they can be aligned with the streamlines in the flow field. In particular, this is useful near surfaces in order to resolve boundary layers. On the other hand, many applications are steady state problems which are determined by the limit of instationary computations. In this context implicit schemes are preferable to explicit schemes in order to use larger time steps when approaching the steady state. In the following we will verify that the adaptive concept can also be applied to an implicit FVS although no rigorous analysis can be provided so far.

7.2.1 Configurations

We consider two configurations which are relevant for the design of an airplane wing.

Bump. This configuration is considered in [RV81]. It is determined by a wall with a bump where the bump is a circular arc with a secant of length $l = 1$ [m] and a thickness of $h = 0.042$ [m], see Figure 7.4. We impose a homogeneous flow field characterized by the free–stream quantities. Here we consider a transonic problem characterized by the Mach number, the temperature and the pressure. Characteristic for this configuration are the kinks in the surface which influence the solution in the neighborhood of these singularities in the geometry. In addition, a local supersonic domain develops caused by the expansion along the bump surface provided the free–stream velocity is sufficiently high. Further downstream the flow is compressed again. This leads to a subsonic domain. According to physical properties the subsonic and the supersonic domain are separated by a compression shock.

Profile. This configuration is determined by a profile of an airfoil, see Figure

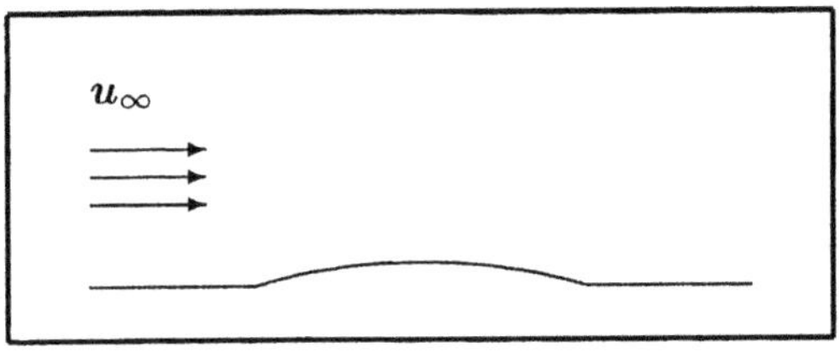

	Units	$\mathbf{u}_\infty$
Ma_∞		0.8
T_∞	K	285
p_∞	Pa	101300

Fig. 7.4. Initial configuration for circular arc bump

7.5. Here we choose the SFB401 cruise configuration according to the BAC 3-11/RES/30/21 profile of Moir, see [Moi94, Pie97]. Characteristic for this problem is the nonsymmetric shape of the profile and the two inflection points of the lower surface where the surface changes from convex to concave and visa versa. Here we only consider axis–symmetric free–stream conditions, i.e., the angle of attack α is put to zero. For sufficiently high free–stream velocities two supersonic domains develop at the upper and lower surface accompanied by a local shock. In addition, a slip line occurs emanating from the trailing edge which is not a straight line due to the nonsymmetric shape of the profile.

7.2.2 Discretization

The computations are performed by means of the new flow solver **QUADFLOW**, see [BBM00, BGMH$^+$01]. This solver merges three basic tools, namely, (i) an implicit FVS on unstructured meshes, (ii) the local multiscale analysis and (iii) a grid generator. The reference implicit FVS is a high–order ENO scheme defined on *unstructured* grids. The core ingredients of this scheme are (i) a genuinely multidimensional linear reconstruction, (ii) the limiter due to Venkatakrishnan [Ven95] and (iii) the HLL Riemann solver

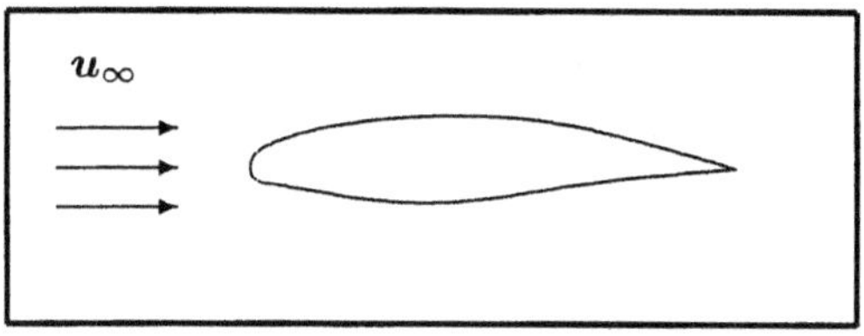

	Units	$\mathbf{u}_\infty$
Ma_∞		0.76
T_∞	K	285
p_∞	Pa	101300

Fig. 7.5. SFB401 cruise configuration

modified according to [TSS94, BLG97]. For the time discretization we apply local time stepping, i.e., for each cell we determine a local step size $\tau_{j,k}$ by a prescribed CFL number c_τ according to

$$\tau_{j,k} \le c_\tau \, |V_{j,k}| \, \left(\max\left\{|\lambda_k(\mathbf{u},\mathbf{n})| \; ; \; \mathbf{u} \in \mathcal{D}, \; \|\mathbf{n}\|_2 = 1\right\}\right)^{-1}.$$

Hence the solution proceeds differently in time for each cell. This is only justified in case of steady state problems. The implicit time discretization is resolved by a Newton–Krylov type method based on a matrix free GMRES and an ILU preconditioner. At the boundaries we apply either slip conditions at the solid wall, i.e., the normal velocity component vanishes, or characteristic boundary conditions for subsonic inflow and outflow boundaries in case of the bump configuration. For the profile we use a point vortex correction according to Thomas and Salas [TS86] at the far field boundary.

The adaptive implicit FVS can be derived analogously to the explicit case. However we have to replace the explicit FVS (4.1) by an implicit FVS that can be written in the form

$$\mathbf{v}_{L,k}^{n+1} + \theta \, \lambda_{L,k} \, \mathsf{B}_{L,k}^{n+1} = \mathbf{v}_{L,k}^{n} - (1-\theta) \, \lambda_{L,k} \, \mathsf{B}_{L,k}^{n}.$$

Here the influence of the implicitness can be controlled by the parameter $\theta \in [0,1]$, e.g. $\theta = 0$ (explicit), $\theta = 1$ (fully implicit). For our computations we always use the fully implicit scheme. Applying the local multiscale transformation to these equations we derive the local evolution equations for the adaptive grid

$$\mathbf{v}_{j,k}^{n+1} + \theta \, \lambda_{j,k} \, \mathsf{B}_{j,k}^{n+1} = \mathbf{v}_{j,k}^{n} - (1-\theta) \, \lambda_{j,k} \, \mathsf{B}_{j,k}^{n}, \quad (j,k) \in \mathcal{G}_{L,\varepsilon}^{n+1},$$

in analogy to (4.11). Since the reference FVS is an unstructured solver, we can perform the local flux evaluation by (4.16) instead of (4.15). Hence the grid need not be locally uniform. In this case it suffices to choose the grading parameter q such that the local transformations are feasible. According to Corollary 4 this is guaranteed provided $q \ge s = 1$. For the computations the threshold value is put to $\varepsilon = 10^{-3}$. Moreover the adaptive grid is updated after N_{adap} time steps instead of each time step in case of an instationary problem.

The underlying grid generator is based on a block–structured concept where the computational domain Ω is decomposed into several blocks, i.e., $\Omega = \bigcup_k \Omega_k$. In Figures 7.6 and 7.7 the decomposition of the computational domains for the above test configurations are sketched.

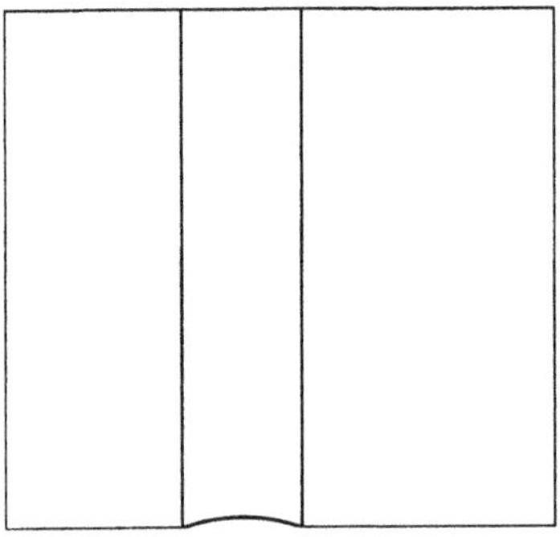

Fig. 7.6. Circular arc bump

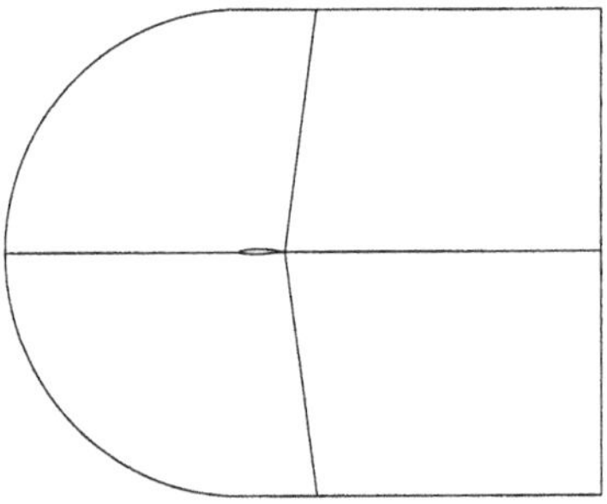

Fig. 7.7. Cruise configuration

In each block Ω_k we apply a structured grid generator. In view of a *sparse* representation of the nested grid hierarchy it is prohibited to store the grid points. To this end, the grid hierarchy is determined by a grid function which maps the parameter domain R onto the block Ω_k, see Section 3.8. In order to provide a fast computation of the grid points as well as to control grid properties, e.g. orthogonality, smoothness, etc., we represent the functions by B–splines, i.e.,

$$\mathbf{x}(\xi) = \sum_{\mathbf{k}} \mathbf{p_k} \prod_{i=1}^{d} N_{k_i}^4(\xi_i),$$

where $\mathbf{p_k} \in \mathsf{R}^d$ denote the control points and N_i^4 the B–spline basis functions of order 4. In general, the number of control points is much smaller than the number of grid points corresponding to the finest resolution level L, see [BM00].

7.2.3 Discussion of Results

For each of the two configurations we perform several computations with increasing number of refinement levels L and a fixed threshold value ε. The characteristic values for the underlying discretizations are summarized in Table 7.7. As initial grid we choose the uniform grid of level 1, because the

Table 7.7. Discretization parameters

	# blocks	$N_{x,0} \times N_{y,0}$	N_T	N_{adap}	c_T
bump	3	5×5	200	20	50
profile	4	5×5	250	25	50

initial data are constant in the entire computational domain according to the free–stream conditions. After N_{adap} time steps we adapt the grid to the current flow field by means of the local multiscale analysis. The steady state is approached after N_T time steps. In Figures A.16, A.17 and A.20, A.21 the local flow field nearby the bump and the profile, respectively, as well as the corresponding adaptive grids are presented. Note, that due to the subsonic free–stream conditions the computational domain has to be chosen sufficiently large in order to avoid perturbations from the far field boundary. For our computations we use 1.5 and 2.5 of the bump span before and behind the bump, respectively, and 2 bump spans in height. For the profile we use about 20 wing spans in each direction. The plots of the numerical results again verify the reliability of the multiscale analysis. All physical meaningful effects have been detected. For both the bump and the profile the shock is recognized. In addition, the stagnation points of the bump configuration which coincide with the kinks in the surface as well as the slip line in the wake of the profile are detected. Note, that in the latter case there is no jump in the pressure field and the normal velocity field. Since no singularity has been imposed by the initial data it is remarkable that these discontinuities evolve with proceeding time. Therefore it is important that details on lower scales may predict significant details on higher scales within the prediction step, see Section 4.1.2.

Table 7.8. Parameter study for the bump

L	N_L	N_G %	C_{MS} [min]	$\|\|\mathbf{r}_L\|\|_1$	$\|\|\mathbf{r}_L\|\|_\infty$
4	19200	9.5	14	2E−9	3E−7
5	96800	2.5	14	2E−8	3E−7
6	387200	1.1	25	1E−8	5E−7
7	1548800	0.5	47	2E−8	2E−6

Again the computational complexity behavior in case of steady state problems is documented in Table 7.8 and 7.9. We note that the number of cells in the final adaptive grid is significantly smaller than that of the uniform grid on the finest resolution level L. In particular, the relative number N_G decreases

Table 7.9. Parameter study for the profile

L	N_L	N_G %	C_{MS} [min]	$\|\|\mathbf{r}_L\|\|_1$	$\|\|\mathbf{r}_L\|\|_\infty$
4	25600	28	30	6E−7	2E−3
5	102400	15	51	3E−7	8E−5
6	409600	8	92	1E−6	2E−5
7	1638400	4	164	6E−7	8E−6

much more rapidly with increasing number of refinement levels for the bump as in case of the test configurations considered in the previous section. This can be explained by the *local* structure of the singularities. For the profile the relative number N_G does not decrease so rapidly with increasing refinement level because the slip line traverses the entire wake whereas the singularities occurring in the bump configuration are local.

In contrast to instationary problems, we have to consider the asymptotic behavior of the solution which is expected to approach a steady state. An indicator is defined by the difference $\mathbf{r}_{j,k} := \mathbf{v}_{j,k}^{n+1} - \mathbf{v}_{j,k}^n$ of the conservative quantities on two successive time levels which frequently referred to as the *(temporal) residual.* In Table 7.8 the *maximal residual* $||\mathbf{r}_L||_\infty$ and the *average residual* $||\mathbf{r}_L||_1$ defined by

$$||\mathbf{r}_L||_\infty := \max_{i=1,\ldots,m} \max_{(j,k)\in\mathcal{G}_{L,\varepsilon}} |r_{j,k,i}|$$

$$||\mathbf{r}_L||_1 := \max_{i=1,\ldots,m} \sum_{(j,k)\in\mathcal{G}_{L,\varepsilon}} (|V_{j,k}|/|\Omega|)|r_{j,k,i}|$$

are listed. For our computations the volume of the computational domain is about $|\Omega| = 10$ (bump) and $|\Omega| = 1271$ (profile). It turns out that the maximal and average residual are reduced by four orders of magnitude. during the computations for both configurations. This is verified by the plots of the temporal variation of the maximal residual for the density, see Figure A.18 and A.22. We observe that between two adaptation steps the residual decreases almost monotonically. However after each grid adaptation the residual significantly increases by several orders of magnitude. This is caused by the thresholding that is performed within the local multiscale analysis. Note, that the new residual after adapting the grid is in the order of the threshold value ε. Nevertheless the residual decreases more strongly after each adaptation step. Approaching the steady state in not only reflected in the decrease of the residuals, but also in the ratio of the number of grid points before and after the adaptation step. In Figure A.19 and A.23 this ratio is presented. Here again we observe an asymptotic behavior. In the limit we approach 1, i.e., the number of grid points is maintained.

A Plots of Numerical Experiments

Fig. A.1. Shock reflection: density contours ($L = 8,\ T = 0$ [s])

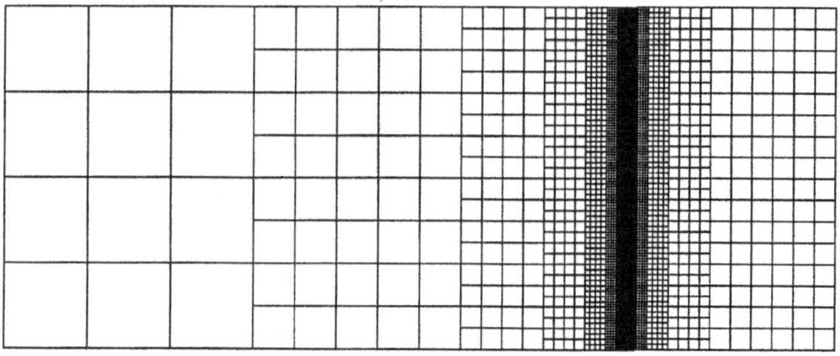

Fig. A.2. Shock reflection: adaptive grid ($L = 8,\ T = 0$ [s])

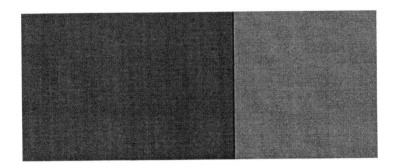

Fig. A.3. Shock reflection: density contours ($L = 8,\ T = 3.2 \times 10^{-4}$ [s])

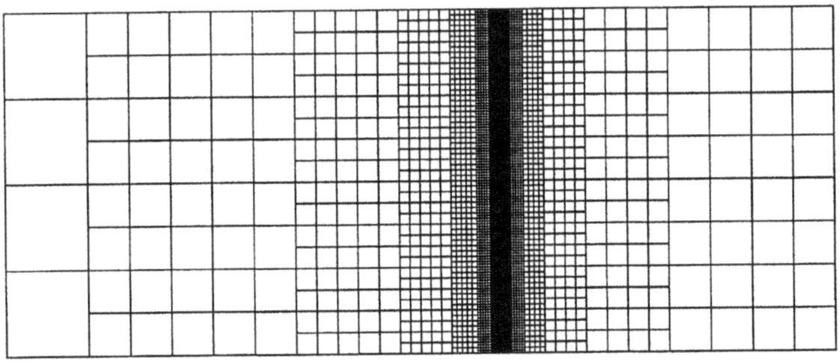

Fig. A.4. Shock reflection: adaptive grid ($L = 8,\ T = 3.2 \times 10^{-4}$ [s])

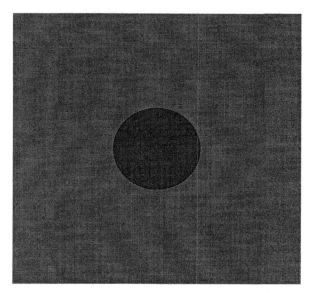

Fig. A.5. Implosion: density contours ($L = 9,\ T = 0$ [s])

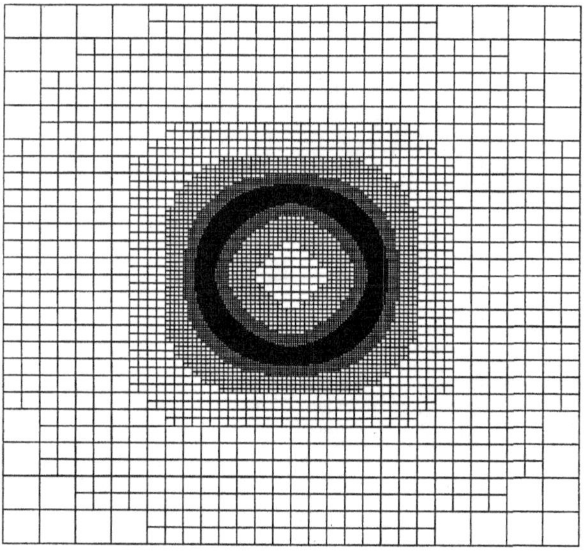

Fig. A.6. Implosion: adaptive grid ($L = 9,\ T = 0$ [s])

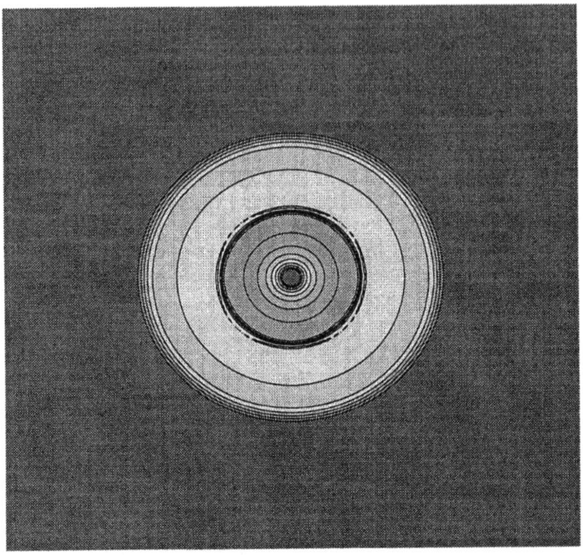

Fig. A.7. Implosion: density contours ($L = 9,\ T = 3.6125 \times 10^{-4}$ [s])

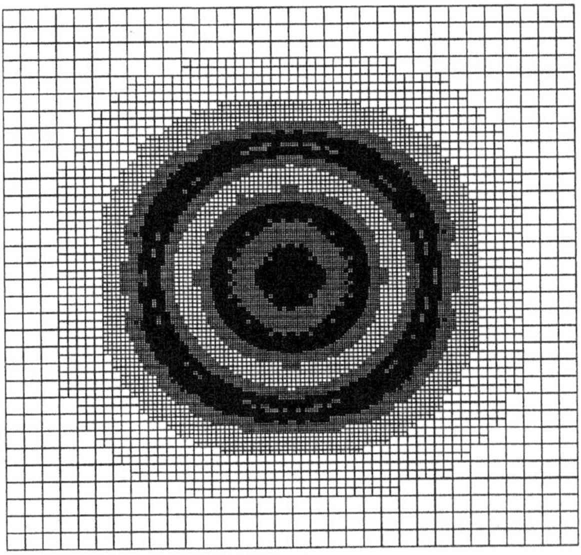

Fig. A.8. Implosion: adaptive grid ($L = 9,\ T = 3.6125 \times 10^{-4}$ [s])

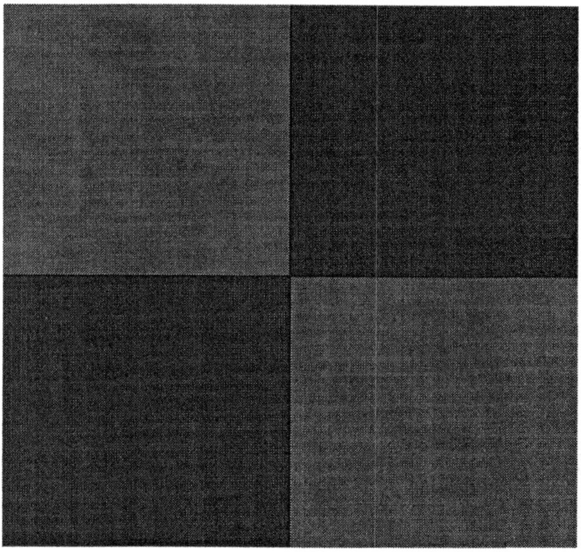

Fig. A.9. Wave interaction: density contours ($L = 6,\ T = 0$ [s])

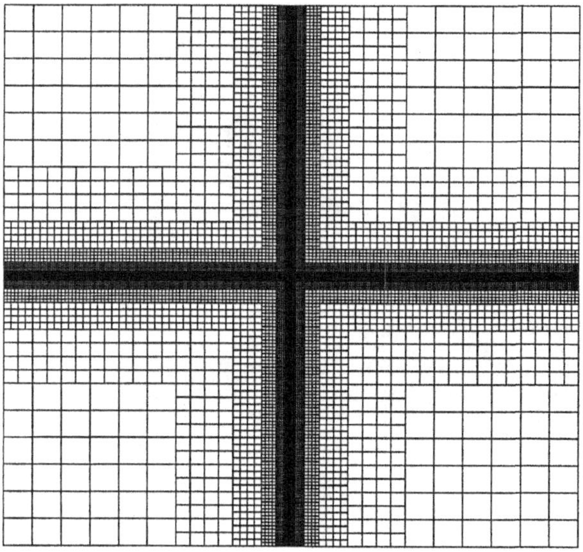

Fig. A.10. Wave interaction: adaptive grid ($L = 6,\ T = 0$ [s])

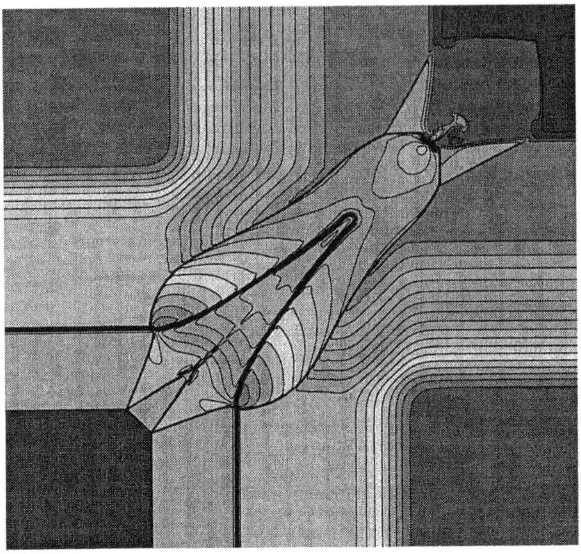

Fig. A.11. Wave interaction: density contours ($L = 6,\ T = 0.15$ [s])

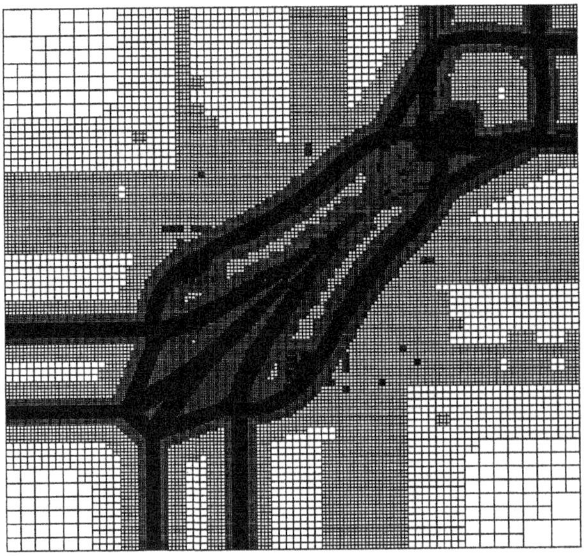

Fig. A.12. Wave interaction: adaptive grid ($L = 6,\ T = 0.15$ [s])

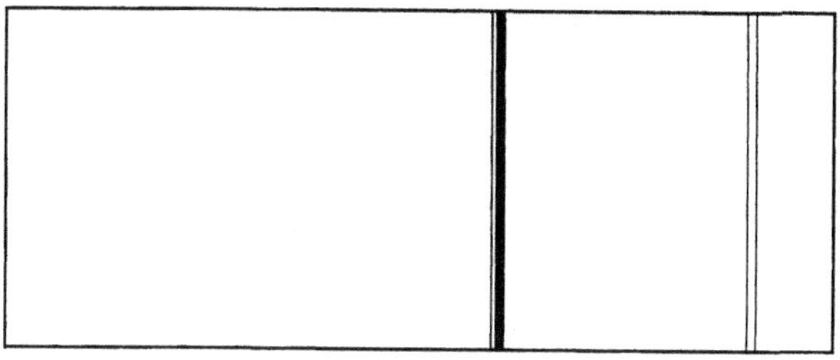

Fig. A.13. Shock reflection: contours of error in density ($L = 7,\ T = 3.2 \times 10^{-4}$ [s])

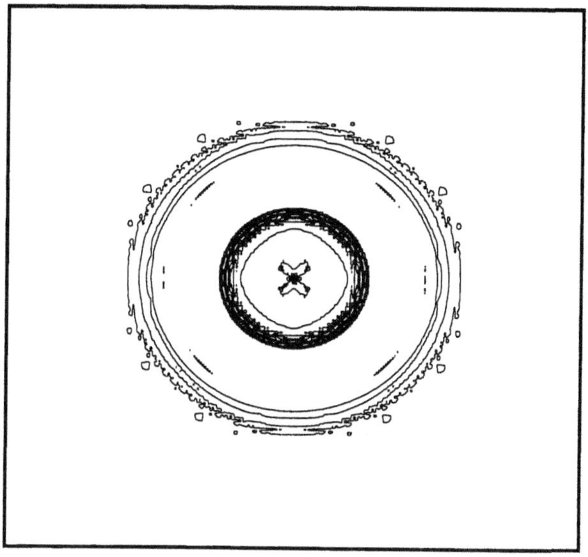

Fig. A.14. Implosion: contours of error in density ($L = 7,\ T = 3.6125 \times 10^{-4}$ [s])

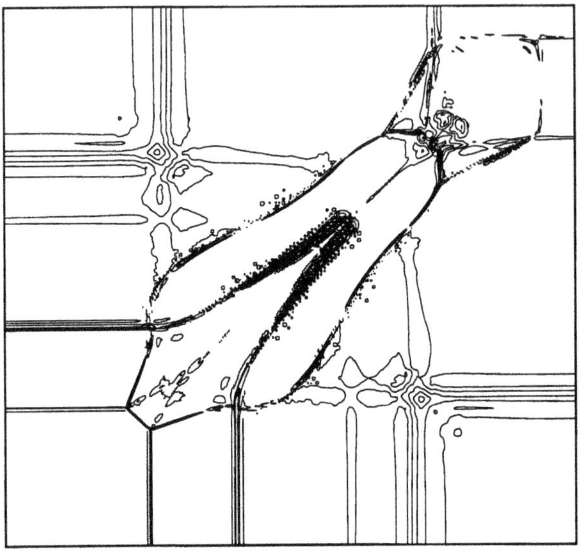

Fig. A.15. Wave interaction: contours of error in density ($L = 5,\ T = 0.15$ [s])

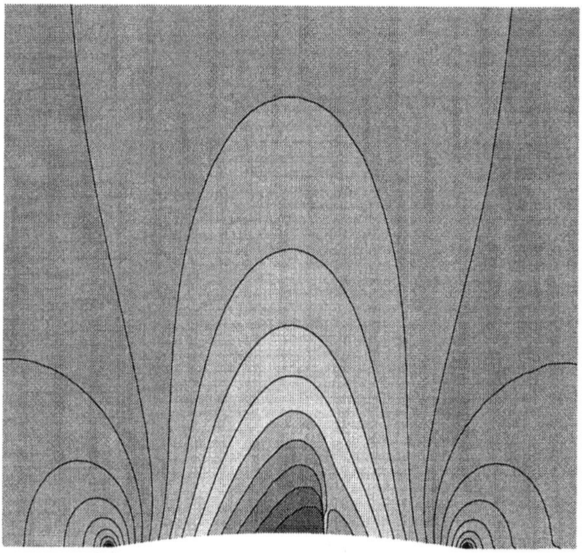

Fig. A.16. Bump: Mach contours ($L = 7$)

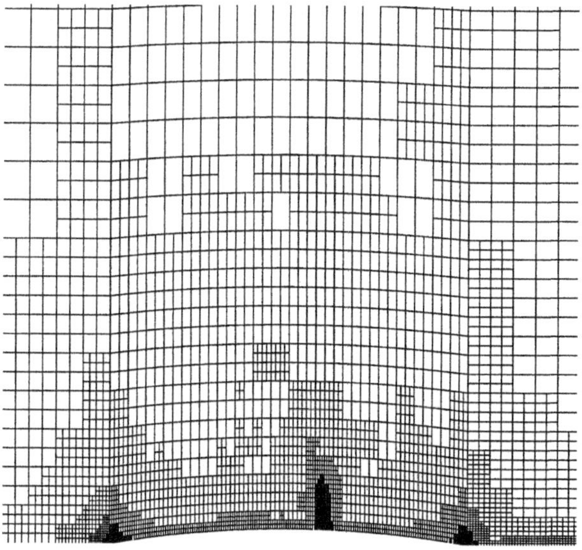

Fig. A.17. Bump: adaptive grid ($L = 7$)

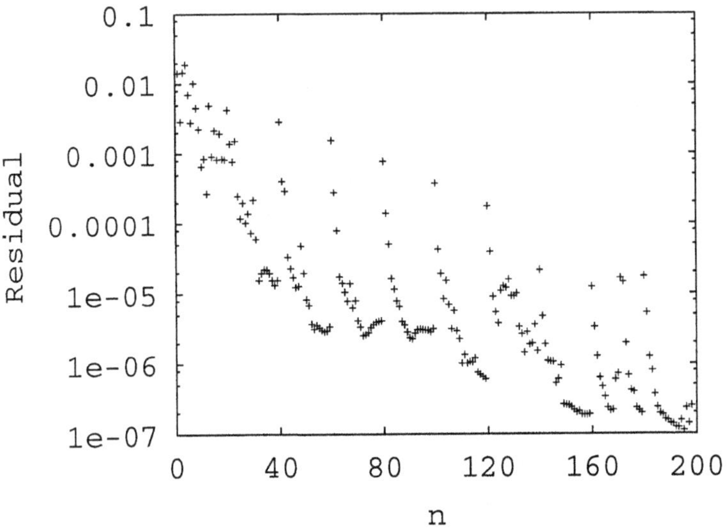

Fig. A.18. Bump: temporal variation of maximal density residual ($L = 7$)

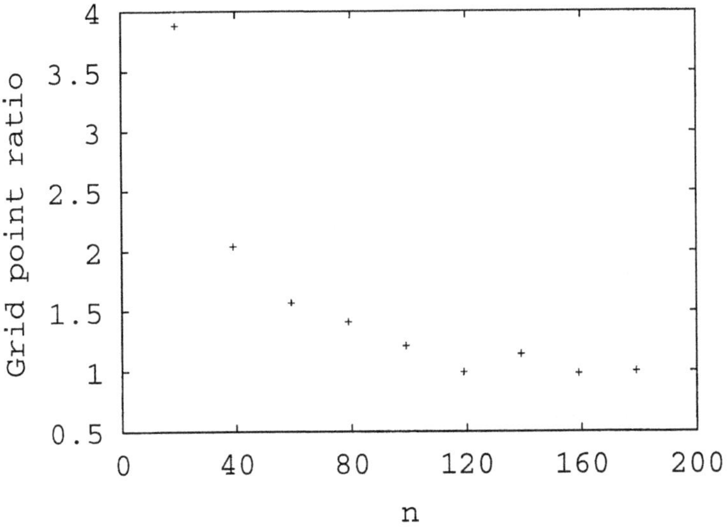

Fig. A.19. Bump: temporal variation of grid size ratio before/after adaptation ($L = 7$)

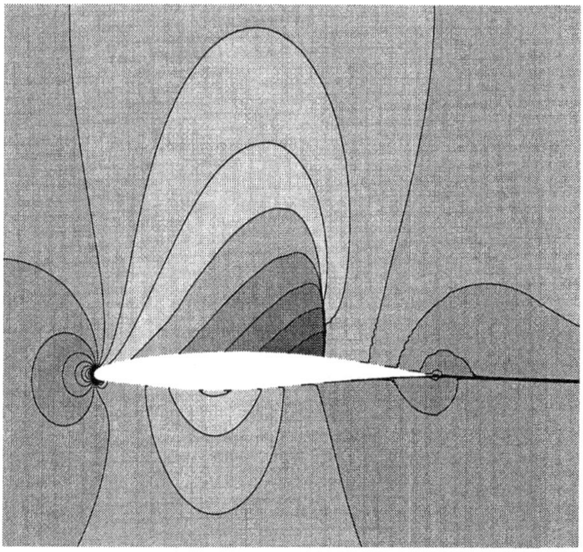

Fig. A.20. Profile: Mach contours ($L = 7$)

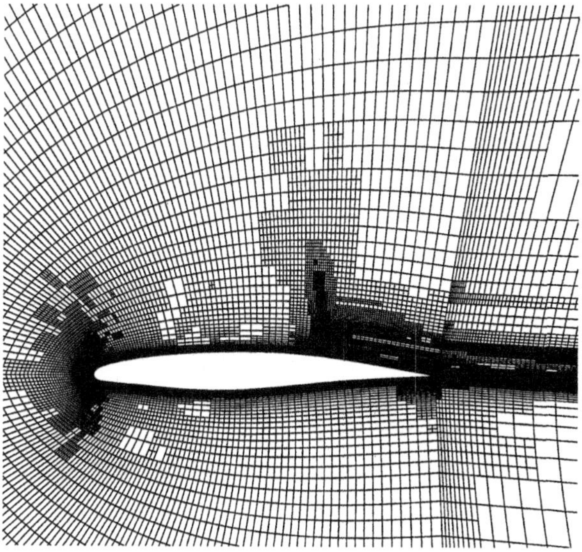

Fig. A.21. Profile: adaptive grid ($L = 7$)

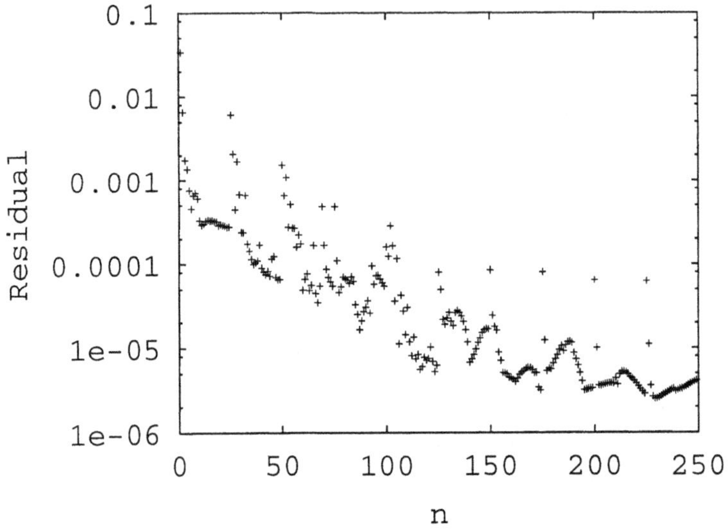

Fig. A.22. Profile: temporal variation of maximal density residual ($L = 7$)

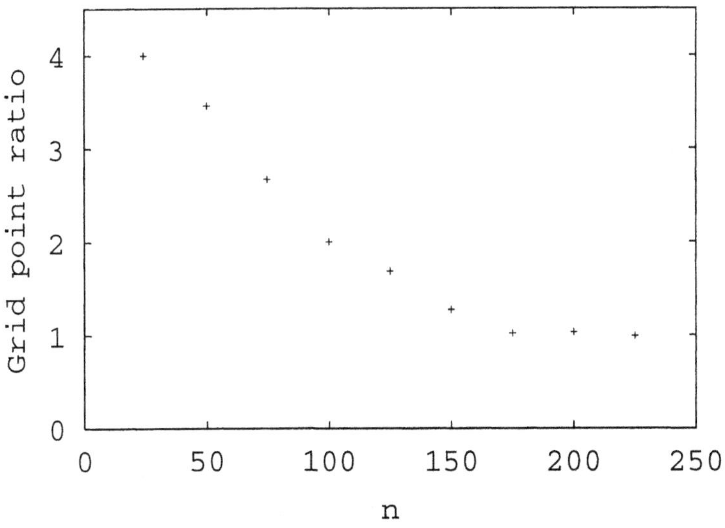

Fig. A.23. Profile: temporal variation of grid size ratio before/after adaptation ($L = 7$)

B The Context of Biorthogonal Wavelets

This section provides a short introduction to (biorthogonal) wavelet theory. Here we focus on some basic aspects that are restricted to our purposes. For a general overview on wavelet theory we recommend standard textbooks such as the recent book by A. Cohen [Coh00].

We start with a general multiresolution analysis following the methodology of W. Dahmen [Dah94, Dah95, Dah96] and collaborators [CDP96]. In particular, we outline the connection between *stability* and *biorthogonal wavelets*. As an example for the general setting we consider wavelets on the real line according to [CDF92].

B.1 General Setting

In the sequel, we consider a Hilbert space[1] $\mathcal{H} = \mathcal{H}(\Omega)$ of functions endowed with an inner product $\langle \cdot, \cdot \rangle_{\mathcal{H}}$ and associated norm $\| \cdot \|_{\mathcal{H}} := \langle \cdot, \cdot \rangle_{\mathcal{H}}^{1/2}$. Then a sequence $\mathcal{S} = \{S_j\}_{j \in \mathbb{N}_0}$ of closed linear subspaces $S_j \subset \mathcal{H}$ is called a *multiresolution* or *multiscale sequence*, if the subspaces are *nested*, i.e.,

$$S_0 \subset S_1 \subset \ldots \subset S_j \subset S_{j+1} \subset \ldots \mathcal{H}$$

and $\mathcal{S}$ is dense in $\mathcal{H}$, i.e.,

$$\mathrm{clos}_{\mathcal{H}} \left(\bigcup_{j \in \mathbb{N}_0} S_j \right) = \mathcal{H}.$$

The nestedness of the multiresolution sequence implies the existence of the *complement* or *wavelet spaces* W_j such that

$$S_{j+1} = S_j \oplus W_j = \ldots = S_0 \oplus \bigoplus_{i=0}^{j} W_i.$$

[1] Note, that stability is best investigated in L^2. Similar results can not be derived in L^1. However, in Sect. 2, the box wavelets and their modification have to be scaled with respect to L^1.

B.1.1 Multiscale Basis

We now assume that the subspaces S_j and W_j are spanned by sets of basis functions

$$\Phi_j := \{\varphi_{j,k}\,;\, k \in I_j\}, \qquad \Psi_j := \{\psi_{j,k}\,;\, k \in J_j\}$$

enumerated by the index sets I_j and J_j, i.e.,

$$S_j = \mathrm{clos}_{\mathcal{H}}\left(\mathrm{span}\{\Phi_j\}\right) =: S(\Phi_j), \qquad W_j = \mathrm{clos}_{\mathcal{H}}\left(\mathrm{span}\{\Psi_j\}\right) =: S(\Psi_j).$$

We refer to Φ_j as the *basis of scaling functions*[2] and Ψ_j as the *wavelet basis.* Here the index sets are either *finite* or at most *countable* sets. In particular, in the finite case we have $N_{j+1} = N_j + K_j$ with $N_j := \#I_j,\ K_j := \#J_j$.

For later use it is convenient to interpret a set of functions Θ_j, e.g., $\Theta_j = \Phi_j$ or $\Theta_j = \Psi_j$, as a column vector with respect to a fixed but unspecified ordering of basis functions. Then it is possible to represent any function $u \in \mathrm{span}\{\Theta_j\}$ by

$$u = \Theta_j^T\,\mathbf{u} := \sum_k u_k\,\theta_{j,k}, \qquad \mathbf{u} \in \mathsf{R}^{\#\Theta_j}.$$

The collection $\{\Theta_j\}_{j\in\mathsf{N}_0}$ is called *uniformly stable*, if there exist constants $c,\ C > 0$ independent of j such that the *Riesz property*

$$c\,\|\mathbf{u}\|_{l^2} \le \|\Theta_j^T\mathbf{u}\|_{\mathcal{H}} \le C\,\|\mathbf{u}\|_{l^2} \tag{B.1}$$

holds for all $\mathbf{u} \in \mathsf{R}^{\#\Theta_j}$ (see [CDP96]). In particular, if Θ_j is an orthonormal system, then the Riesz property holds with "=" instead of "$\le$" and $c = C = 1$. The Riesz property (B.1) with uniform constants ensures the uniqueness of the expansion of u with respect to the basis Θ_j. If the basis Θ_j satisfies the Riesz property then it is called a *Riesz basis* for the space $S(\Theta_j)$.

Keeping our application in mind, we confine the discussion to uniformly stable Riesz bases, i.e., $\{\Phi_j\}_{j\in\mathsf{N}_0}$ and $\{\Psi_j\}_{j\in\mathsf{N}_0}$ satisfy the Riesz property (B.1). In the context of principal shift–invariant spaces S_j, i.e., if $\mathcal{H}$ is a Hilbert space over R^d and $\mathcal{S}$ satisfies

$$f \in S_j \qquad \text{if and only if} \qquad f(\cdot - 2^{-j}\mathbf{k}) \in S_j, \quad \mathbf{k} \in \mathsf{Z}^d,$$

the notion of a multiscale sequence coincides with the concept of Mallat's *multiresolution analysis* [Mal89]. An example is discussed in Sect. B.2. From the nestedness of $\mathcal{S}$ and the uniform stability of the Riesz bases, we conclude the existence of bounded linear operators $\mathsf{M}_{j,0}$ and $\mathsf{M}_{j,1}$ such that the two–scale relations

$$\Phi_j^T = \Phi_{j+1}^T \mathsf{M}_{j,0}, \qquad \Psi_j^T = \Phi_{j+1}^T \mathsf{M}_{j,1} \tag{B.2}$$

or componentwise

[2] Here the notion of *scaling function* is used in a generalized sense.

$$\varphi_{j,k} = \sum_{r \in I_{j+1}} m_{r,k}^{j,0}\, \varphi_{j+1,r}, \qquad \psi_{j,k} = \sum_{r \in I_{j+1}} m_{r,k}^{j,1}\, \varphi_{j+1,r}$$

hold. These relations can be interpreted as a change of bases realized by the matrices

$$\mathsf{M}_{j,0} = (m_{r,k}^{j,0})_{r \in I_{j+1}, k \in I_j}, \qquad \mathsf{M}_{j,1} = (m_{r,k}^{j,1})_{r \in I_{j+1}, k \in J_j},$$

provided that the inverse of the composed matrix $\mathsf{M}_j := (\mathsf{M}_{j,0}, \mathsf{M}_{j,1})$ exists. Note, that in practice a two–scale relation for the scaling functions Φ_j is known. Then the task is to determine $\mathsf{M}_{j,1}$ such that the composed matrix M_j is invertible. If the Φ_j are uniformly stable the existence of the inverse $\mathsf{G}_j := \mathsf{M}_j^{-1}$ together with the uniform boundedness of M_j and G_j, i.e.,

$$||\mathsf{M}_j||_{[l_2(I_{j+1}), l_2(I_j \cup J_j)]}, \; ||\mathsf{G}_j||_{[l_2(I_j \cup J_j), l_2(I_{j+1})]} = \mathcal{O}(1), \qquad j \to \infty. \tag{B.3}$$

is equivalent to the uniform stability of $\Phi_j \cup \Psi_j$, i.e., $\Phi_j \cup \Psi_j$ is also a Riesz basis for $S(\Phi_{j+1}) = S(\Phi_j \cup \Psi_j)$, cf. [CDP96]. According to the blocking of M_j it is convenient to represent the inverse by $\mathsf{G}_j := (\mathsf{G}_{j,0}^T, \mathsf{G}_{j,1}^T)^T$ where the blocks are determined by

$$\mathsf{G}_{j,0} = (g_{r,k}^{j,0})_{r \in I_{j+1}, k \in I_j}, \qquad \mathsf{G}_{j,1} = (g_{r,k}^{j,1})_{r \in I_{j+1}, k \in J_j}.$$

In particular, these blocks fulfill

$$\mathsf{M}_{j,0}\, \mathsf{G}_{j,0} + \mathsf{M}_{j,1}\, \mathsf{G}_{j,1} = \mathsf{I}, \qquad \mathsf{G}_{j,i}\, \mathsf{M}_{j,i'} = \delta_{i,i'} \mathsf{I}, \quad i, i' \in \{0, 1\}. \tag{B.4}$$

By means of these identities we deduce the two–scale equation

$$\Phi_{j+1}^T = \Phi_j^T \mathsf{G}_{j,0} + \Psi_j^T \mathsf{G}_{j,1} \tag{B.5}$$

or componentwise

$$\varphi_{j+1,k} = \sum_{r \in I_j} g_{r,k}^{j,0}\, \varphi_{j,r} + \sum_{r \in J_j} g_{r,k}^{j,1}\, \psi_{j,r}.$$

Note, that we can equivalently switch between the two–scale representations (B.2) and (B.5) where we employ the identities (B.4).

B.1.2 Stable Completion

From the discussion above we conclude that the task of determining a basis for the complement spaces is directly connected to the task of determining an appropriate matrix $\mathsf{M}_{j,1}$, i.e., the problem has been transferred from functional analysis to linear algebra. This is summarized in the following definition.

Definition 10. *(Uniformly stable completion)*
Any matrix $\mathsf{M}_{j,1} \in [\mathsf{R}^{N_{j+1}}, \mathsf{R}^{K_j}]$ *satisfying the relations (B.4) and (B.3) is called a* stable completion *of* $\mathsf{M}_{j,0}$.

Although the problem of constructing a wavelet basis has now become a pure algebraic problem it is nevertheless not easy to solve. However, the set of all stable completions of $\mathsf{M}_{j,0}$ can be characterized if just one stable completion $\check{\mathsf{M}}_{j,1}$ of $\mathsf{M}_{j,0}$ is known.

Theorem 8. *(Change of stable completion [CDP96])*
Let $\check{\mathsf{M}}_{j,1}$ *be a fixed stable completion of* $\mathsf{M}_{j,0}$ *and* $\mathsf{L}_j \in [\mathsf{R}^{K_j}, \mathsf{R}^{N_j}]$, $\mathsf{K}_j \in [\mathsf{R}^{K_j}, \mathsf{R}^{K_j}]$ *uniformly bounded linear operators such that* K_j *is regular. Then*

$$\mathsf{M}_{j,1} = \check{\mathsf{M}}_{j,1}\mathsf{K}_j + \mathsf{M}_{j,0}\mathsf{L}_j \tag{B.6}$$

is also a stable completion of $\mathsf{M}_{j,0}$. *In particular, the inverse of* $\mathsf{M}_j = (\mathsf{M}_{j,0}, \mathsf{M}_{j,1})$ *is determined by* $\mathsf{G}_j^T = (\mathsf{G}_{j,0}, \mathsf{G}_{j,1})$ *with*

$$\mathsf{G}_{j,0} = \check{\mathsf{G}}_{j,0} - \mathsf{L}_j\mathsf{K}_j^{-1}\check{\mathsf{G}}_{j,1}, \qquad \mathsf{G}_{j,1} = \mathsf{K}_j^{-1}\check{\mathsf{G}}_{j,1}. \tag{B.7}$$

The reverse statement is summarized in the following theorem.

Theorem 9. *(Characterization of stable completions [CDP96])*
Let $\mathsf{M}_{j,1}$, $\check{\mathsf{M}}_{j,1}$ *be two stable completions of* $\mathsf{M}_{j,0}$. *Then there exist uniformly bounded linear operators* $\mathsf{L}_j \in [\mathsf{R}^{K_j}, \mathsf{R}^{N_j}]$, $\mathsf{K}_j \in [\mathsf{R}^{K_j}, \mathsf{R}^{K_j}]$ *with inverse* $\mathsf{K}_j^{-1} \in [\mathsf{R}^{K_j}, \mathsf{R}^{K_j}]$ *such that (B.6) and (B.7) hold.*

In terms of the new stable completion, the modified wavelet basis can be represented in terms of the old wavelet basis $\check{\Psi}_j$ employing the matrices L_j and K_j, i.e.,

$$\Psi_j^T = \Phi_{j+1}^T\mathsf{M}_{j,1} = \Phi_{j+1}^T\check{\mathsf{M}}_{j,1}\mathsf{K}_j + \Phi_{j+1}^T\mathsf{M}_{j,0}\mathsf{L}_j = \check{\Psi}_j^T\mathsf{K}_j + \Phi_j^T\mathsf{L}_j. \tag{B.8}$$

Note, that the basis Φ_j of scaling functions is not changed, since $\check{\mathsf{M}}_{j,0} = \mathsf{M}_{j,0}$. Only the wavelet basis $\check{\Psi}_j$ is modified according to a linear combination of scaling functions on level j and an appropriate weighting.

B.1.3 Multiscale Transformation

So far we have only considered the relations between two successive levels. Now we are interested in the multiscale case. For this purpose, we introduce the *multiscale basis*

$$\Psi^{(L)} := \bigcup_{j=-1}^{L-1} \Psi_j, \qquad \Psi_{-1} := \Phi_0.$$

By means of this basis an element $u \in S_L$ can be expanded with respect to the bases Φ_L and $\Psi^{(L)}$, i.e.,

$$u = \Phi_L^T\mathbf{u}_L = (\Psi^{(L)})^T\mathbf{d}^{(L)} = \sum_{j=-1}^{L-1} \Psi_j^T\mathbf{d}_j \tag{B.9}$$

with the multiscale coefficients

$$\mathbf{d}^{(L)} = (\mathbf{d}_{-1}^T, \mathbf{d}_0^T, \dots, \mathbf{d}_{L-1})^T, \qquad \mathbf{d}_{-1} := \mathbf{u}_0. \tag{B.10}$$

In the sequel the two different basis expansions are referred to as *single-scale representation* and *wavelet or multiscale representation*, respectively. If the systems $\{\Phi_j\}_{j\in\mathbb{N}_0}$ and $\{\Phi_j \cup \Psi_j\}_{j\in\mathbb{N}_0}$ are uniformly stable Riesz bases, then there is a bijective linear operator $\mathcal{M}_L : \mathbb{R}^{N_L} \to \mathbb{R}^{N_L}$ which performs the change of bases between Φ_L and $\Psi^{(L)}$, see (B.9), i.e.,

$$\mathbf{d}^{(L)} = \mathcal{M}_L \mathbf{u}_L \quad \text{and} \quad \mathbf{u}_L = \mathcal{M}_L^{-1} \mathbf{d}^{(L)}.$$

The operator $\mathcal{M}_L$ is called the *multiscale transformation* and $\mathcal{M}_L^{-1}$ the *inverse multiscale transformation.* In order to obtain an explicit representation of $\mathcal{M}_L$ and its inverse we conclude from the two-scale relations (B.2) and (B.5) that the coefficients of two successive levels are related by

$$\mathbf{u}_{j+1} = \mathsf{M}_j \begin{pmatrix} \mathbf{u}_j \\ \mathbf{d}_j \end{pmatrix}, \qquad \begin{pmatrix} \mathbf{u}_j \\ \mathbf{d}_j \end{pmatrix} = \mathsf{G}_j \mathbf{u}_{j+1}. \tag{B.11}$$

Iterative application of (B.11) yields the pyramid schemes in Figures B.1 and B.2 which are similar to the up- and down-sampling in signal processing. In matrix notation, the multiscale transformation is given by

$$\mathcal{M}_L = \overline{\mathsf{G}}_0 \cdots \overline{\mathsf{G}}_{L-1}, \qquad \mathcal{M}_L^{-1} = \overline{\mathsf{M}}_{L-1} \cdots \overline{\mathsf{M}}_0 \tag{B.12}$$

with the matrices

$$\overline{\mathsf{M}}_j = \begin{pmatrix} \mathsf{M}_j & 0 \\ 0 & \mathsf{I}_j \end{pmatrix}, \qquad \overline{\mathsf{G}}_j = \begin{pmatrix} \mathsf{G}_j & 0 \\ 0 & \mathsf{I}_j \end{pmatrix}, \qquad \mathsf{I}_j := \mathsf{I}_{N_L - N_{j+1}}.$$

Analogously to the two-scale setting we are again interested in the stability of the transformation and the uniqueness of the basis expansion. These correspond to the uniform Riesz basis property (B.1) for the multiscale basis $\Psi^{(L)}$, $L \in \mathbb{N}_0$. Investigations in [CDP96] show that this property is equivalent to the uniform boundedness of the multiscale transformations, i.e.,

$$\|\mathcal{M}_L\|_{[l_2(I_L), l_2(J^{(L)})]}, \ \|\mathcal{M}_L^{-1}\|_{[l_2(J^{(L)}), l_2(I_L)]} = \mathcal{O}(1), \qquad L \to \infty,$$

where the index set $J^{(L)} := \bigcup_{j=-1}^{L-1} J_j$, $J_{-1} := I_0$, corresponds to the multiscale basis $\Psi^{(L)}$. This condition is equivalent to the existence of a *dual Riesz basis* for $\mathcal{H}$.

Theorem 10. *(Dahmen [Dah94])*
If the bases $\{\Phi_j\}_{j\in\mathbb{N}_0}$ are uniformly stable, then the multiscale transformation is uniformly stable if and only if $\Psi := \Phi_0 \cup \{\Psi_j\}_{j\in\mathbb{N}_0}$ is a Riesz basis for $\mathcal{H}$. In this case, there is another Riesz basis $\tilde{\Psi} := \tilde{\Phi}_0 \cup \{\tilde{\Psi}_j\}_{j\in\mathbb{N}_0}$ of $\mathcal{H}$ which is biorthogonal to Ψ, i.e., $\langle \Psi, \tilde{\Psi} \rangle = \mathsf{I}$ and $u \in \mathcal{H}$ has the unique expansion

$$u = \sum_{k\in I_0} \langle u, \tilde{\varphi}_{0,k} \rangle_{\mathcal{H}} \varphi_{0,k} + \sum_{j\in\mathbb{N}_0} \sum_{k\in J_j} \langle u, \tilde{\psi}_{j,k} \rangle_{\mathcal{H}} \psi_{j,k}. \tag{B.13}$$

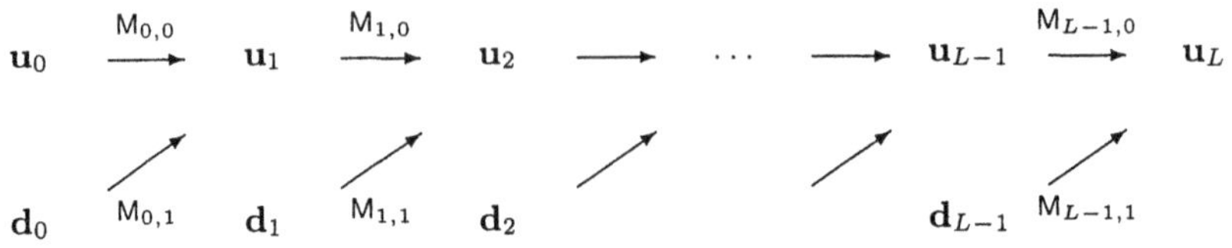

Fig. B.1. Pyramid scheme of the multiscale transformation

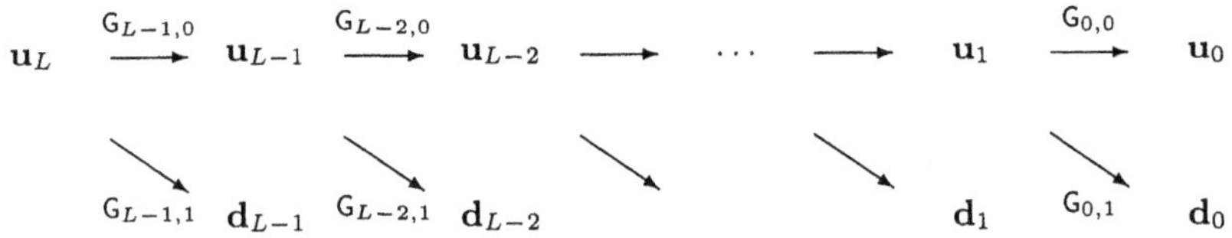

Fig. B.2. Pyramid scheme of the inverse multiscale transformation

We emphasize that the uniform stability of the multiscale bases $\{\tilde{\Phi}_0 \cup \tilde{\Psi}_0 \cup \ldots \cup \tilde{\Psi}_L\}_{L\in\mathbb{N}}$ can only be expected to hold in a Hilbert space, e.g. $\mathcal{H} = L^2$. In the present work, however, the relevant topology here is ultimately determined by L^1. In particular, we make essential use of the uniform stability of the levelwise bases $\{\tilde{\Phi}_j\}_{j\in\mathbb{N}_0}$ and $\{\tilde{\Phi}_j \cup \tilde{\Psi}_j\}_{j\in\mathbb{N}_0}$ in L^1.

In the sequel, the bases Φ_j and Ψ_j are always referred to as *primal scaling functions* and *primal wavelets*, respectively. Analogously, the bases $\tilde{\Phi}_j$ and $\tilde{\Psi}_j$ are called the *dual scaling functions* and *dual wavelets*, respectively. This terminology is often used to express that the functions under consideration are expanded in the primal bases while bounded linear functionals are represented in the dual basis. In particular, if the bases $\{\tilde{\Phi}_j\}_{j\in\mathbb{N}_0}$ and $\{\tilde{\Phi}_j \cup \tilde{\Psi}_j\}_{j\in\mathbb{N}_0}$ are uniformly stable and there are two–scale relations of the form (B.2) and (B.5) for the dual bases with appropriate matrices $\widetilde{\mathsf{M}}_{j,i}$, $\widetilde{\mathsf{G}}_{j,i}$, $i = 0, 1$, then the following relations hold between the primal and dual setting, see [CDP96],

$$\widetilde{\mathsf{M}}_{j,0} = \mathsf{G}_{j,0}^T,\ \widetilde{\mathsf{M}}_{j,1} = \mathsf{G}_{j,1}^T, \tag{B.14}$$

$$\langle \Phi_j, \tilde{\Psi}_j \rangle_{\mathcal{H}} = \mathbf{0},\ \langle \Psi_j, \tilde{\Phi}_j \rangle_{\mathcal{H}} = \mathbf{0}, \tag{B.15}$$

$$S_j \perp \widetilde{W}_j := \operatorname{span}\{\tilde{\Psi}_j\},\ \tilde{S}_j := \operatorname{span}\{\tilde{\Phi}_j\} \perp W_j. \tag{B.16}$$

An example for the bases Φ_j, Ψ_j on the primal side and $\widetilde{\Phi}_j, \widetilde{\Psi}_j$ on the dual side is given in Sect. B.2 within the context of principal shift–invariant spaces. Note that the dual multiresolution sequence $\widetilde{\mathcal{S}} = \{\widetilde{S}_j\}_{j\in\mathbb{N}_0}$ is also a dense sequence of closed linear subspaces in $\mathcal{H}$ and satisfies relations analogously to (B.2) and (B.5).

Finally, we would like to remark that in the literature the concept of primal and dual wavelets is referred to as the concept of *biorthogonal wavelets.* For the construction of biorthogonal wavelets in one dimension see [CDF92] ($\Omega = \mathsf{R}$) and [DKU99] ($\Omega = [a, b] \subset \mathsf{R}$).

B.2 Biorthogonal Wavelets of the Box Function

In this section, we consider an example that concerns the class of *principal shift–invariant spaces* where the multiscale sequence $\mathcal{S}$ coincides with the classical conception of a *multiresolution analysis* [Dau88, Mal89, Mey90, JM91, Chu92, CDF92, Dau92]. Here, the Hilbert space is $\mathcal{H} = L^2(\mathsf{R}^d)$, and the subspaces S_j are generated by means of one single function $\varphi \in L^2(\mathsf{R}^d)$. Nestedness and stability imply that φ is *refinable*, i.e., there is a sequence $\mathbf{a} = \{a_\mathbf{k}\}_{\mathbf{k}\in\mathsf{Z}^d} \in l_2(\mathsf{Z}^d)$ denoted as *mask* of φ such that

$$\varphi(\mathbf{x}) = \sum_{\mathbf{k}\in\mathsf{Z}^d} a_\mathbf{k}\, \varphi(2\,\mathbf{x} - \mathbf{k}), \qquad \text{a. e.,} \tag{B.17}$$

is satisfied. We start with spaces

$$S_j = \operatorname{clos}_{L^2(\mathsf{R}^d)} \left(\operatorname{span}\, \{\varphi_{j,\mathbf{k}} \mid \mathbf{k} \in \mathsf{Z}^d\}\right), \qquad j \in \mathsf{Z}, \tag{B.18}$$

with a countable basis consisting of the functions

$$\varphi_{j,\mathbf{k}}(\mathbf{x}) := 2^{jd/2}\varphi(2^j\mathbf{x} - \mathbf{k}), \qquad j \in \mathsf{Z},\ \mathbf{k} \in \mathsf{Z}^d. \tag{B.19}$$

Here the normalization factor $2^{jd/2}$ is chosen such that

$$||\varphi_{j,\mathbf{k}}||_{L^2(\mathsf{R}^d)} = ||\varphi||_{L^2(\mathsf{R}^d)} = 1.$$

B.2.1 Haar Wavelets

As a special case, we consider the *box function* or *B–spline of order 1* in one spatial dimension

$$\varphi(x) := \chi_{[0,1)}(x) = \begin{cases} 1, & 0 \le x < 1 \\ 0, & \text{elsewhere} \end{cases}, \tag{B.20}$$

which obviously satisfies

$$\varphi(x) = \varphi(2x) + \varphi(2x-1), \tag{B.21}$$

i. e., $a_0 = a_1 = 1$ and $a_k = 0$ otherwise in (B.17). Haar [Haa10] established already in 1910 that the system

$$\psi_{j,k} := 2^{j/2}\psi(2^j x - k), \qquad j, k \in \mathsf{Z},$$

for

$$\psi(x) := \varphi(2x) - \varphi(2x-1) = \begin{cases} 1, & 0 \le x < 1/2 \\ -1, & 1/2 \le x < 1 \\ 0, & \text{elsewhere} \end{cases}, \tag{B.22}$$

is an orthonormal basis for $L^2(\mathsf{R})$ which is called *Haar basis.* Since orthonormal bases are, in particular, Riesz bases (with $\varphi_{j,k} = \widetilde{\varphi}_{j,k}$, $\psi_{j,k} = \widetilde{\psi}_{j,k}$), the multiscale transformation is uniformly stable. One particular advantage of the Haar basis for computations is that the system

$$\{\varphi_{0,0}\} \cup \{\psi_{j,k} \mid j \ge 0,\ k = 0, 1, \ldots, 2^j - 1\}$$

is also an orthonormal basis for $L^2([0,1])$ due to the small compact support of φ and ψ. Another advantage of the Haar basis is that the multiscale transformations are very efficiently executable which is also caused by the minimal compact support. For the interval $[0,1]$, the matrices M_j and G_j are banded with

$$\mathsf{M}_{j,i} = (\mathsf{G}_{j,i})^T = \frac{1}{\sqrt{2}} \begin{pmatrix} 1 & 0 & \cdots & \cdots & 0 \\ (-1)^i & 0 & & & \vdots \\ 0 & 1 & & & \\ 0 & (-1)^i & & & \\ \vdots & \ddots & \ddots & \ddots & \vdots \\ & & & 1 & 0 \\ & & & (-1)^i & 0 \\ \vdots & & & 0 & 1 \\ 0 & \cdots & \cdots & 0 & (-1)^i \end{pmatrix} \tag{B.23}$$

for $i = 0, 1$. Unfortunately, the Haar wavelet has only one vanishing moment, i. e.,

$$\int_{\mathsf{R}} \psi(x)\, dx = 0, \qquad \int_{\mathsf{R}} x^m\, \psi(x)\, dx \neq 0,\ m \in \mathsf{N}.$$

B.2.2 Biorthogonal Wavelets on the Real Line

In view of a high compression of the multiscale coefficients, we are interested in another function ψ that has more vanishing moments. In [CDF92], a more suitable Riesz basis is constructed. First, it is proved that for each fixed odd number $\widetilde{N} = 2m + 1 > 0$, there is a dual function $\widetilde{\varphi} \in L^2(\mathsf{R}) = {}_{1,\widetilde{N}}\widetilde{\varphi} \in L^2(\mathsf{R})$ which is refinable, has compact support $[1 - \widetilde{N}, \widetilde{N}]$ and satisfies

$$\langle \varphi,\ \widetilde{\varphi}(\cdot - k)\rangle_{L^2(\mathsf{R})} = \delta_{k,0}, \quad k \in \mathsf{Z},$$

such that the sets

$$\mathbf{\Phi}_j = \{\varphi_{j,k} \mid k \in \mathsf{Z}\}, \qquad \widetilde{\mathbf{\Phi}}_j = \{\widetilde{\varphi}_{j,k} \mid k \in \mathsf{Z}\}$$

are biorthogonal to each other, i. e.,

$$\left\langle \mathbf{\Phi}_j,\, \widetilde{\mathbf{\Phi}}_j \right\rangle = \mathsf{I},$$

where $\widetilde{\varphi}_{j,k}$ is defined analogously to $\varphi_{j,k}$. The mask $\widetilde{\mathbf{a}} = \{\widetilde{a}_k\}_{k\in\mathsf{Z}}$ of $\widetilde{\varphi}$ satisfying the refinement relation

$$\widetilde{\varphi}(x) = \sum_{k\in\mathsf{Z}} \widetilde{a}_k\, \widetilde{\varphi}(2\,x - k), \qquad \text{a. e.,} \tag{B.24}$$

has only non–vanishing elements for $k = 1 - \widetilde{N}, \ldots, \widetilde{N}$ and can explicitly be determined as coefficients of the Laurent polynomial

$$\sum_{k=1-\widetilde{N}}^{\widetilde{N}} \widetilde{a}_k z^k = (4z)^{-m}(z+1)^{2m+1} \sum_{n=0}^{m} \binom{m+n}{n} (-4z)^{-n}(z-1)^{2n}. \tag{B.25}$$

Additionally, it is shown in [CDF92] that the functions

$$\psi(x) = \sum_{k=1-\widetilde{N}}^{\widetilde{N}} b_k\, \varphi(2\,x - k), \tag{B.26}$$

$$\widetilde{\psi}(x) = \widetilde{\varphi}(2\,x) - \widetilde{\varphi}(2\,x - 1) \tag{B.27}$$

with $\mathbf{b} = \{b_k\}_{k\in\mathsf{Z}}$ and $b_k := (-1)^{k+1}\widetilde{a}_{1-k}$ generate the complement spaces

$$\begin{aligned} W_j &= S(\mathbf{\Psi} j) = \operatorname{clos}_{L^2(\mathsf{R})} \left(\operatorname{span} \{\psi_{j,k} \mid k \in \mathsf{Z}\}\right), \\ \widetilde{W}_j &= S(\widetilde{\mathbf{\Psi}} j) = \operatorname{clos}_{L^2(\mathsf{R})} \left(\operatorname{span} \{\widetilde{\psi}_{j,k} \mid k \in \mathsf{Z}\}\right), \end{aligned}$$

where $\psi_{j,k}$ and $\widetilde{\psi}_{j,k}$ are defined analogously to $\varphi_{j,k}$. The multiscale matrices M_j and G_j are determined by (B.21), (B.24), (B.26), (B.27) as follows: Since $\mathsf{M}_{j,0}$ and $\mathsf{G}_{j,1}$ are bi–infinite continuations of (B.23), the masks $\widetilde{\mathbf{a}}/\sqrt{2}$ and $\mathbf{b}/\sqrt{2}$ build the columns of $(\mathsf{G}_{j,0})^T$ and $\mathsf{M}_{j,1}$, respectively, i. e.,

$$\frac{1}{\sqrt{2}}\begin{pmatrix} & \vdots & & \\ \ddots & 0 & \vdots & \\ \ddots & \widetilde{a}_{-\widetilde{N}+1} & 0 & \\ & \widetilde{a}_{-\widetilde{N}+2} & 0 & \\ & \widetilde{a}_{-\widetilde{N}+3} & \widetilde{a}_{-\widetilde{N}+1} & \ddots \\ \ddots & \vdots & \vdots & \ddots \\ \ddots & \widetilde{a}_{\widetilde{N}} & \widetilde{a}_{\widetilde{N}-2} & \\ & 0 & \widetilde{a}_{\widetilde{N}-1} & \\ & 0 & \widetilde{a}_{\widetilde{N}} & \\ & \vdots & 0 & \ddots \\ & & \vdots & \ddots \end{pmatrix}, \qquad \frac{1}{\sqrt{2}}\begin{pmatrix} & \vdots & & \\ \ddots & 0 & \vdots & \\ \ddots & b_{-\widetilde{N}+1} & 0 & \\ & b_{-\widetilde{N}+2} & 0 & \\ & b_{-\widetilde{N}+3} & b_{-\widetilde{N}+1} & \ddots \\ \ddots & \vdots & \vdots & \ddots \\ \ddots & b_{\widetilde{N}} & b_{\widetilde{N}-2} & \\ & 0 & b_{\widetilde{N}-1} & \\ & 0 & b_{\widetilde{N}} & \\ & \vdots & 0 & \ddots \\ & & \vdots & \ddots \end{pmatrix}.$$

A further result of [CDF92] is that $\widetilde{\varphi}$ is exact of order $\widetilde{N}$, i.e.,

$$x^r = \sum_{k \in \mathsf{Z}} \langle (\cdot)^r, \varphi_{j,k} \rangle \, \widetilde{\varphi}_{j,k}, \qquad r = 0, 1, \ldots, \widetilde{N} - 1,$$

such that $\mathsf{P}_{\widetilde{N}-1} \subset \widetilde{S}_j$ for $j \in \mathsf{Z}$. This controls the approximation properties of the dual sequence $\widetilde{\mathcal{S}}$ and, additionally, it implies together with relation (B.16) that

$$\int_{\mathsf{R}} x^m \, \psi(x) \, dx = 0, \qquad m = 0, 1, \ldots, \widetilde{N} - 1,$$

i. e., ψ has $\widetilde{N}$ vanishing moments. Thus, for $\widetilde{N} \geq 3$, the functions $\widetilde{\varphi}$ are better candidates for generators of $\widetilde{\mathcal{S}}$ than the orthogonal choice φ which corresponds to $\widetilde{N} = 1$.

Finally, we present the primal and dual scaling functions and wavelets, respectively, corresponding to $\widetilde{N} = 1, 3, 5$, cf. Figures B.3, B.4, B.5. We observe that the primal functions are always discontinuous whereas the functions on the dual side become the more regular the higher $\widetilde{N}$ is chosen. In our applications, see Sect. 2, the roles of the primal and dual side are reversed. Moreover, these functions are scaled with respect to L^1 and L^∞ on the dual and primal side, respectively, instead of L^2 in the Hilbert space setting considered here.

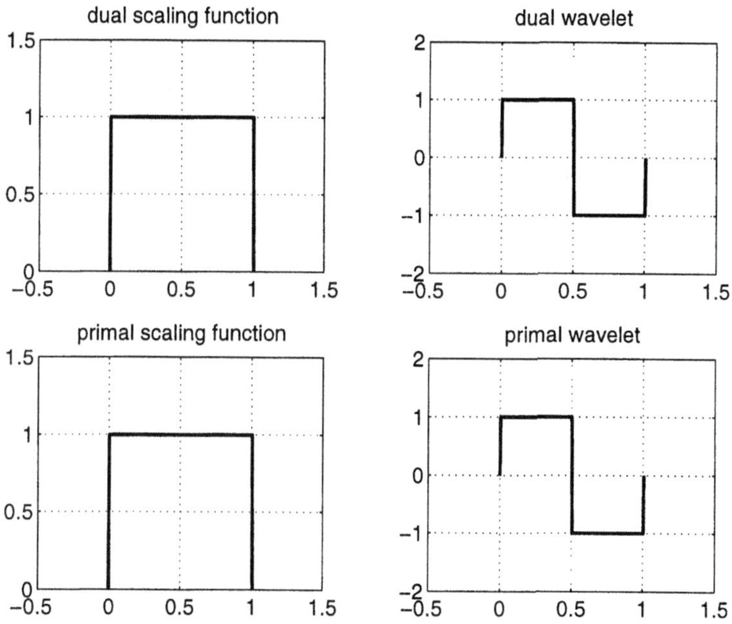

Fig. B.3. Biorthogonal wavelets and scaling functions for $\widetilde{N} = 1$

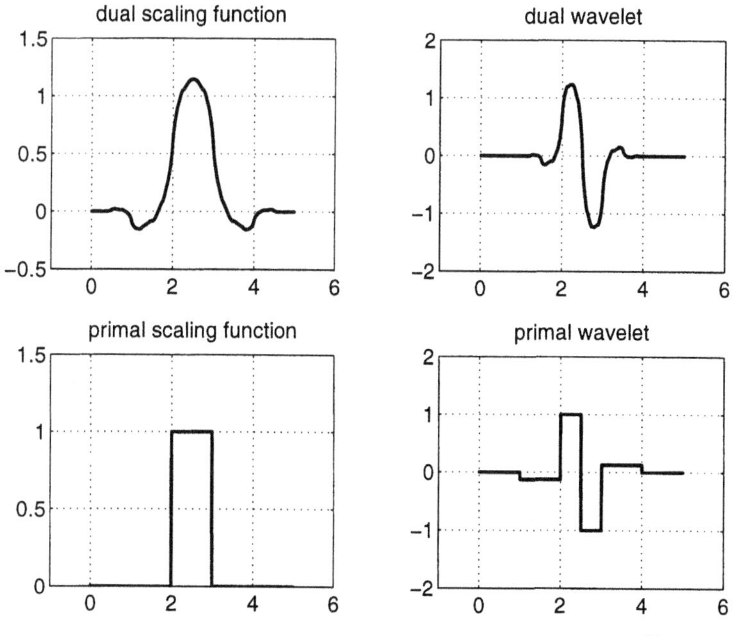

Fig. B.4. Biorthogonal wavelets and scaling functions for $\widetilde{N} = 3$

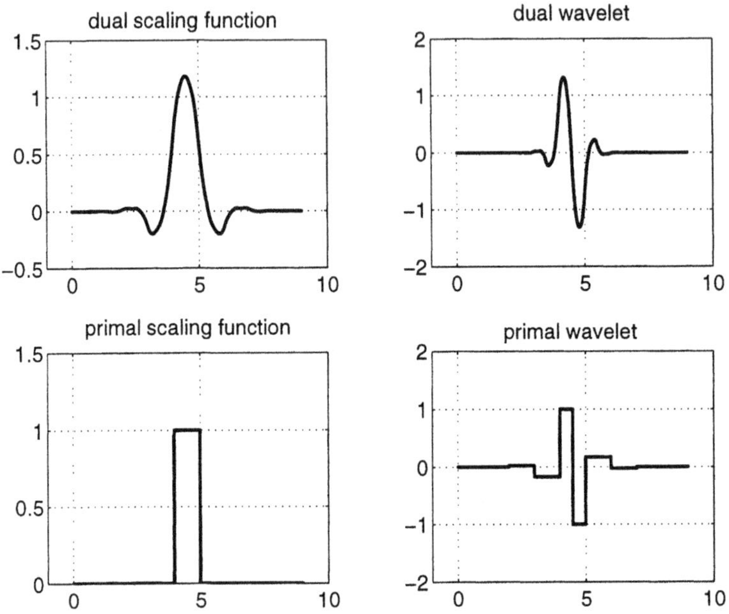

Fig. B.5. Biorthogonal wavelets and scaling functions for $\widetilde{N} = 5$

References

[Abg91] R. Abgrall. Design of an Essentially Non–Oscillatory Reconstruction Procedure on Finite–Element Type Meshes. ICASE Report 91–84, NASA, 1991.

[Abg97] R. Abgrall. Multiresolution analysis on unstructured meshes: Applications to CFD. In Chetverushkin and al., editors, *Experimentation, modelling and computation in flow, turbulence and combustion*. John Wiley & Sons, 1997.

[ADH98] F. Arandiga, R. Donat, and A. Harten. Multiresolution based on weighted averages of the hat function I: Linear reconstruction techniques. *SIAM J. Numer. Anal.*, 36(1):160–203, 1998.

[ADH99] F. Arandiga, R. Donat, and A. Harten. Multiresolution based on weighted averages of the hat function II: Non-linear reconstruction techniques. *SIAM J. Sci. Comput.*, 20(3):1053–1093, 1999.

[BBB73] C. Bardos, D. Brézis, and H. Brézis. Perturbations singulières et prolongements maximaux d'operteurs positifs. *Arch. Ration. Mech. Anal.*, 53:69–100, 1973.

[BBM00] J. Ballmann, F. Bramkamp, and S. Müller. Development of a flow solver employing local adaptation based on multiscale analysis on B–spline grids. In *Proceedings of 8th Annual Conference of the CFD Society of Canada, June, 11 to 13, 2000 (Montreal)*, 2000.

[BGMH+01] F. Bramkamp, B. Gottschlich-Müller, M. Hesse, Ph. Lamby, S. Müller, J. Ballmann, Brakhage K.-H. and Dahmen W. H-adaptive Multiscale Schemes for the Compressible Navier–Stokes Equations — Polyhedral Discretization, Data Compression and Mesh Generation. IGPM–Report 207, RWTH Aachen, 2001. Accepted for publication in Numerical Notes on Fluid Mechanics.

[BH97] B. Bihari and A. Harten. Multiresolution schemes for the numerical solution of 2–D conservation laws I. *SIAM J. Sci. Comput.*, 18(2):315–354, 1997.

[BLG97] P. Batten, M.A. Leschziner, and U.C. Goldberg. Average-state Jacobians and implicit methods for compressible viscous and turbulent flows. *J. Comp. Physics*, 137:38–78, 1997.

[BLN79] C. Bardos, A.Y. LeRoux, and J.C. Nedelec. First order quasilinear equations with boundary conditions. *Comm. Partial Differential Equations*, 4:1017–1034, 1979.

[BM00] K.-H. Brakhage and S. Müller. Algebraic–hyperbolic grid generation with precise control of intersection of angles. *J. Num. Meth. in Fluids*, 33(1):89–123, 2000.

[BO84] M. Berger and J. Oliger. Adaptive mesh refinement for hyperbolic partial differential equations. *J. Comp. Physics*, 53:484–512, 1984.

[BR96] R. Becker and R. Rannacher. A feed-back approach to error control in finite element methods: Basic analysis and examples. *East-West J. Num. Math.*, 4:237–264, 1996.

[CCL94] B. Cockburn, F. Coquel, and P. LeFloch. An error estimate for finite volume methods for multidimensional conservation laws. *Math. Comp.*, 63:77–103, 1994.

[CD01] G. Chiavassa and R. Donat. Point value multiresolution for 2D compressible flows. *SIAM J. Sci. Comput.*, 23(3):805–823, 2001.

[CDD01] A. Cohen, W. Dahmen, and R. DeVore. Adaptive wavelet methods for elliptic operator equations – convergence rates. *Math. Comp.*, 70:27–75, 2001.

[CDF92] A. Cohen, I. Daubechies, and J. Feauveau. Bi–orthogonal bases of compactly supported wavelets. *Comm. Pure Appl. Math.*, 45:485–560, 1992.

[CDKP00] A. Cohen, N. Dyn, S.M. Kaber, and M. Postel. Multiresolution finite volume schemes on triangles. *J. Comp. Physics*, 161:264–286, 2000.

[CDM91] A. Cavaretta, W. Dahmen, and C.A. Micchelli. Stationary subdivision. *Mem. Am. Math. Soc*, 453:186 p, 1991.

[CDP96] J.M. Carnicer, W. Dahmen, and J.M. Peña. Local decomposition of refinable spaces and wavelets. *Appl. Comput. Harmon. Anal.*, 3:127–153, 1996.

[CF48] R. Courant and K.O. Friedrichs. *Supersonic Flow and Shock Waves*. Pure and Applied Mathematics. Interscience Publishers, New York, 1948.

[CG96] B. Cockburn and P.A. Gremaud. Error estimates for finite element methods for scalar conservation laws. *SIAM J. Numer. Anal.*, 33:522–554, 1996.

[Chu92] C.K. Chui. *An Introduction to Wavelets*. Academic Press, Boston, 1992.

[CKMP01] A. Cohen, S.M. Kaber, S. Müller, and M. Postel. Fully Adaptive Multiresolution Finite Volume Schemes for Conservation Laws. *Math. Comp.*, 2001. posted on December 5, 2001, PII S 0025-5718(01)01391-6 (to appear in print).

[CKP02] A. Cohen, S.M. Kaber, , and M. Postel. Multiresolution Analysis on Triangles: Application to Gas Dynamics. In G. Warnecke and H. Freistühler, editors, *Hyperbolic Problems: Theory, Numerics, Applications*, pages 257–266. Birkhäuser, 2002.

[CKS99] B. Cockburn, G. Karniadakis, and C. Shu. The Development of Discontinuous Galerkin Methods. In C. Shu B. Cockburn, G. Karniadakis, editor, *Discontinuous Galerkin Methods: Theory, Computation and Applications*, pages 1–50. Springer, 1999.

[CM79] A.J. Chorin and J.E. Marsden. *A Mathematical Introduction to Fluid Mechanics*. Springer Verlag, Berlin, 1979.

[CM80] M.G. Crandall and A. Majda. Monotone difference approximations for scalar conservation laws. *Math. Comp.*, 34:1–24, 1980.

[Coh00] A. Cohen. Wavelet Methods in Numerical Analysis. In P.G. Ciarlet and J.L. Lions, editors, *Handbook of Numerical Analysis*, Handbook of Numerical Analysis, pages 417–711. Elsevier, Amsterdam, 2000.

[Daf00] C.M. Dafermos. *Hyperbolic Conservation Laws in Continuum Physics*. Grundlehren der Mathematischen Wissenschaften. Springer Verlag, Berlin, 2000.

[Dah94] W. Dahmen. Some remarks on multiscale transformations, stability and biorthogonality. In P.J. Laurent, A. Le Méhauté, and L.L. Schumaker, editors, *Wavelets, Images and Surface Fitting*, pages 157–188. A K Peters, Wellesley, 1994.

[Dah95] W. Dahmen. Multiscale analysis, approximation and interpolation spaces. In C.K. Chui and L.L. Schumaker, editors, *Approximation Theory VIII*, pages 47–88. World Scientific Publishing Co., 1995.

[Dah96] W. Dahmen. Stability of multiscale transformations. *J. Fourier Anal. Appl.*, 4:341–362, 1996.

[Dau88] I. Daubechies. Orthonormal Bases of Compactly Supported Wavelets. *Comm. Pure Appl. Math.*, XLI(41):909–996, 1988.

[Dau92] I. Daubechies. Ten Lectures on Wavelets. SIAM, Philadelphia, 1992.

[DeV98] R. DeVore. Nonlinear approximation. ACTA Numerica, pages 51–151. Cambridge University Press, 1998.

[DGM00] W. Dahmen, B. Gottschlich–Müller, and S. Müller. Multiresolution schemes for conservation laws. *Numer. Math.*, 88(3):399–443, 2000.

[DKU99] W. Dahmen, A. Kunoth, and K. Urban. Biorthogonal spline-wavelets on the interval – stability and moment conditions. *Appl. Comput. Harmon. Anal.*, 6:132–196, 1999.

[DL93] R.A. DeVore and G.G. Lorentz. *Constructive Approximation*. Grundlehren der Mathematischen Wissenschaften. Springer, Berlin, 1993.

[DMS02] W. Dahmen, S. Müller, and Th. Schlinkmann. On an adaptive multigrid solver for convection-dominated problems. *SIAM J. Sci. Comput.*, 23(3):781–804, 2002.

[DSX00] W. Dahmen, R. Schneider, and Y. Xu. Nonlinear functionals of wavelet expansions – adaptive reconstruction and fast evaluation. *Numer. Math.*, 86(1):49–101, 2000.

[Dyn92] N. Dyn. Subdivision algorithms in computer–aided geometric design. In W.A. Light, editor, *Advances in Numerical Analysis*, Oxford, 1992. Clarendon Press.

[EEHJ95] K. Eriksson, D. Estep, P. Hanspo, and C. Johnson. Introduction to adaptive methods for differential equations. ACTA Numerica, pages 105–158. Cambridge University Press, 1995.

[EJ91] K. Eriksson and C. Johnson. Adaptive finite element methods for parabolic problems. I. A linear model problem. *SIAM J. Numer. Anal.*, 28:43–77, 1991.

[EJ95] K. Eriksson and C. Johnson. Adaptive finite element methods for parabolic problems. IV. Nonlinear problems. *SIAM J. Numer. Anal.*, 32:1729–1749, 1995.

[FL71] K.O. Friedrichs and P.D. Lax. Systems of conservation equations with convex extensions. In *Proc. Nat. Acad. Sci. USA*, volume 68, No. 8, pages 1686–1688, August 1971.

[FM72] A. Fischer and D.P. Marsden. The Einstein evolution equations as a first order quasi–linear symmetric hyperbolic system. *Comm. Math. Physics*, 28:1–38, 1972.

[Gel59] I. Gelfand. Some problems in the theory of quasilinear equations. *Usp. Mat. Nauk*, 14:87, 1959. Am. Math. Soc. Transl., Ser. 2, 29, 295 (1963).

[GM99a] B. Gottschlich–Müller and S. Müller. Adaptive finite volume schemes for conservation laws based on local multiresolution techniques. In M. Fey and R. Jeltsch, editors, *Hyperbolic Problems: Theory, Numerics, Applications*, pages 385–394. Birkhäuser, 1999.

[GM99b] B. Gottschlich–Müller and S. Müller. On multi–scale concepts for multi–dimensional conservation laws. In W. Hackbusch and G. Wittum, editors, *Numerical Treatment of Multi–scale Problems*, pages 119–133. Vieweg, 1999.

[God59] S.K. Godunov. A finite difference method for the numerical computation and discontinuous solutions of the equations of fluid dynamics. *Math. Sb*, 47:271–295, 1959.

[Got98] B. Gottschlich–Müller. *Multiscale Schemes for Conservation Laws.* PhD thesis, RWTH Aachen, 1998.

[GV96] G.H. Golub and C.F. Van Loan. *Matrix Computations.* John–Hopkins University Press, Baltimore, 3rd edition, 1996.

[Haa10] A. Haar. Zur Theorie der orthogonalen Funktionen–Systeme. *Math. Ann*, 69:331–371, 1910.

[Har93a] A. Harten. Discrete multi–resolution analysis and generalized wavelets. *J. Appl. Num. Math.*, 12:153–193, 1993.

[Har93b] A. Harten. Multiresolution Representation of Data, I. Preliminary Report. UCLA CAM Report 93–13, 1993.

[Har94] A. Harten. Adaptive multiresolution schemes for shock computations. *J. Comp. Phys.*, 115:319–338, 1994.

[Har95] A. Harten. Multiresolution algorithms for the numerical solution of hyperbolic conservation laws. *Comm. Pure Appl. Math.*, 48(12):1305–1342, 1995.

[Har96] A. Harten. Multiresolution representation of data: A general framework. *SIAM J. Numer. Anal.*, 33(3):1205–1256, 1996.

[HC91] A. Harten and S.R. Chakravarthy. Multi–dimensional ENO Schemes for General Geometries. ICASE Report 91–76, NASA, 1991.

[HEOC87] A. Harten, B. Engquist, S. Osher, and S.R. Chakravarthy. Uniformly high order accurate essentially non–oscillatory schemes III. *J. Comp. Phys.*, 71:231–303, 1987.

[HR02] R. Hartmann and R. Rannacher. Adaptive FE-methods for conservation laws. In G. Warnecke and H. Freistühler, editors, *Hyperbolic Problems: Theory, Numerics, Applications*, pages 495–504. Birkhäuser, 2002.

[JM91] R.-Q. Jia and C.A. Micchelli. Using the refinement equations for the construction of pre-wavelets II: Powers of two. In P.J. Laurent, A. Le Méhauté, and L.L. Schumaker, editors, *Curves and Surfaces.* Academic Press, New York, 1991.

[Joh87] C. Johnson. *Numerical Solution of Partial Differential Equations by the Finite Element Method.* Cambridge University Press, 1987.

[JS95] C. Johnson and A. Szepessy. Adaptive finite element methods for conservation laws based on a posteriori error estimates. *Comm. Pure Appl. Math.*, 48:199–234, 1995.

[KL89] H.O. Kreis and J. Lorenz. *Initial Boundary Value Problems and the Navier–Stokes Equations.* Academic Press, London, 1989.

[KO99] D. Kröner and M. Ohlberger. A posteriori error estimates for upwind finite volume schemes for nonlinear conservation laws in multi dimensions. *Math. Comp.*, 69(229):25–39, 1999.

[Kre70] H.O. Kreis. Initial boundary value problems for hyperbolic systems. *Comm. Pure Appl. Math.*, 23:276–288, 1970.

[Krö97] D. Kröner. *Numerical Schemes for Conservation Laws.* Advances in Numerical Mathematics. Wiley–Teubner, 1997.

[Kru70] S.N. Kruzhkov. First order quasilinear equations with several space variables. *Math. USSR Sb.*, 10:217–243, 1970.

[Kuz67] N.N. Kuznetsov. The weak solution of the Cauchy problem for a multi-dimensional quasilinear equation. *Mat. Zametki*, 2:401–410, 1967. In Russian.

[Lax54] P.D. Lax. Weak solutions of nonlinear hyperbolic equations and their computation. *Comm. Pure Appl. Math.*, 7:159–193, 1954.

[LeV92] R.J. LeVeque. *Numerical Methods for Conservation Laws.* Birkhäuser Verlag, Basel, 2nd edition, 1992.

[Liu75] T.-P. Liu. The Riemann problem for general systems of conservation laws. *J. Diff. Eqns.*, 18:218–234, 1975.

[Liu76] T.-P. Liu. Shock waves in the nonisentropic gas flow. *J. Diff. Eqns.*, 22:442–452, 1976.

[LW60] P.D. Lax and B. Wendroff. Systems of conservation laws. *Comm. Pure Appl. Math.*, 13:217–237, 1960.

[Mal89] S. Mallat. Multiresolution Approximations and Wavelet Orthonormal Bases of $L^2(\mathsf{R})$. *Trans. Amer. Math. Soc.*, 315(1):69–87, 1989.

[Mey90] Y. Meyer. *Ondelettes et Opérateurs*, volume II. Hermann Editeurs des Sciences et des Arts, Paris, 1990.

[Moi94] I.R.M. Moir. Measurements on a Two–Dimensional Aerofoil with High–Lift Devices. *AGARD-AR-303*, 2:58–59, 1994.

[Mül93] S. Müller. *Erweiterung von ENO–Verfahren auf zwei Raumdimensionen und Anwendung auf hypersonische Stauspunktprobleme.* PhD thesis, RWTH Aachen, 1993.

[Mül02] S. Müller. Adaptive multiresolution schemes. In B. Herbin, editor, *Finite Volumes for Complex Applications.* Hermes Science, Paris, 2002.

[MV00] S. Müller and A. Voss. A Manual for the Template Class Library `igpm_t_lib`. IGPM–Report 197, RWTH Aachen, 2000.

[Noe96] S. Noelle. A note on entropy inequalities and error estimates for higher–order accurate finite volume schemes on irregular families of grids. *Math. Comp.*, 65:1155–1163, 1996.

[OW96] T. Ottmann and P. Widmayer. *Algorithmen und Datenstrukturen.* Spektrum Akademischer Verlag, Heidelberg, Berlin, Oxford, 3rd edition, 1996.

[Pie97] C. Pieper. Definition der Referenzprofile. Interner Bericht, Institut für Luft- und Raumfahrt, 1997.

[Roe81] P. Roe. Approximate Riemann solvers, parameter vectors, and difference schemes. *J. Comp. Phys.*, 43:357–372, 1981.

[RS02] O. Roussel and K. Schneider. A fully adaptive multiresolution scheme for 3D reaction–diffusion equations. In B. Herbin, editor, *Finite Volumes for Complex Applications.* Hermes Science, Paris, 2002.

[RV81] A. Rizzi and H. Viviand. *Numerical methods for the computation of inviscid transonic flows with socck waves*, volume 3 of *Notes on Numerical Fluid Mechanics*. Vieweg, Braunschweig, 1981.

[SB80] J. Stoer and R. Bulirsch. *Introduction to Numerical Analysis*. Springer, New York, 1980.

[Sch98] C. Schwab. *p- and hp-Finite Element Methods*. Clarendon Press, Oxford, 1998.

[SH97] E. Süli and P. Houston. Finite element methods for hyperbolic problems: aposteriori analysis and adaptivity. In I.S. Duff(ed.) et al., editor, *The state of the art in numerical analysis*, pages 441–471. Clarendon Press, Oxford, 1997.

[Son95] Th. Sonar. Multivariate Rekonstruktionsvorschrift zur numerischen Berechnung hyperbolischer Erhaltungsgleichungen, 1995. Habilitation thesis.

[SSF00] F. Schröder–Pander, T. Sonar, and O. Friedrich. Generalized multiresolution analysis on unstructured grids. *Numer. Math.*, 86:685–715, 2000.

[Str97] B. Stroustrup. *The C++ Programming Language*. Addison–Wesley, Reading, 1997.

[Swe98] W. Sweldens. The lifting scheme: A construction of second generation wavelets. *SIAM J. Math. Anal.*, 29(2):511–546, 1998.

[Tad91] E. Tadmor. Local error estimates for discontinuous solutions of nonlinear hyperbolic equations. *SIAM J. Numer. Anal.*, 28:891–906, 1991.

[Tor97] E.F. Toro. *Riemann Solvers and Numerical Methods for Fluid Dynamics*. Springer Verlag, Berlin Heidelberg, 1997.

[TS86] J.L. Thomas and M.D. Salas. Far–field boundary conditions for transonic lifting solutions to the Euler equations. *AIAA J.*, 24:1074–1080, 1986.

[TSS94] E.F. Toro, M. Spruce, and W. Speares. Restoration of the contact surface in the HLL Riemann solver. *Shock Waves*, 4:25–34, 1994.

[Ven95] V. Venkatakrishnan. Convergence to steady state solutions of the Euler equations on unstructured grids with limiters. *J. Comp. Physics*, 118:120–130, 1995.

[Ver95] R. Verfürth. *A Review of A Posteriori Estimation and Adaptive Mesh-Refinement Techniques*. Advances in Numerical Mathematics. Wiley/Teubner, New York-Stuttgart, 1995.

[Vil94] J.P. Vila. Convergence and error estimates in finite volume schemes for general multidimensional scalar conservation laws. I. Explicit monotone schemes. *R.A.I.R.O. Anal. Numer.*, 28:267–295, 1994.

List of Figures

List of Tables

Notation

Conservation Laws

d	number of spatial directions
m	number of conservation laws
$\Omega \subset \mathsf{R}^d$	computational domain
$\check{D} \subset \mathsf{R}^m$	space of admissible states
$t \in \mathsf{R}_0^+$	time variable
$\mathbf{x} \in \Omega$	space variable
$V \subset \Omega$	control volume
∂V	boundary of control volume
$\mathbf{n} \in R^d$	outer normal to ∂V
$T \in \mathsf{R}^+$	integration time
$\mathbf{u} \in \check{D}$	vector of conserved quantities
$\mathbf{f}_i \in \mathsf{R}^m$	flux in ith coordinate direction
$\mathbf{f}_\mathbf{n} \in \mathsf{R}^m$	flux in normal direction $\mathbf{n}$
$\mathsf{A}(\mathbf{u}, \mathbf{n})$	Jacobian of $\mathbf{f}_\mathbf{n}$ at state $\mathbf{u}$
$\mathbf{r}_i(\mathbf{u}, \mathbf{n})$	ith right eigenvector of $\mathsf{A}(\mathbf{u}, \mathbf{n})$
$\mathbf{l}_i(\mathbf{u}, \mathbf{n})$	ith left eigenvector of $\mathsf{A}(\mathbf{u}, \mathbf{n})$
$\lambda(\mathbf{u}, \mathbf{n})$	ith eigenvalue (characteristic speed) of $\mathsf{A}(\mathbf{u}, \mathbf{n})$

Finite Volume Scheme

$n \in \mathsf{N}_0$	time step
$k \in \mathsf{N}_0$	position in space
t_n	nth time step
τ	temporal step size
V_k	kth cell of discretization of Ω
$\lvert V_k \rvert$	volume of cell V_k
∂V_k	boundary of cell V_k
$\Gamma_{k,l}$	interface between cell V_k and V_l

$\|\Gamma_{k,l}\|$	volume of interface $\Gamma_{k,l}$
$\hat{\mathbf{u}}_k^n$	cell average of exact solution $\mathbf{u}$ at time t_n in cell V_k
$\hat{\mathbf{f}}_{k,l}^n$	average flux across the interface $\Gamma_{k,l}$
$\mathbf{n}_{k,l}$	outer normal of cell V_k at the interface $\Gamma_{k,l}$
$\mathbf{v}_k^n$	approximation of $\hat{\mathbf{u}}_k^n$
$\mathsf{F}_{k,l}^n$	approximation of $\hat{\mathbf{f}}_{k,l}^n$

Grid Hierarchy

j	refinement level
L	maximal number of refinement levels
k	position on an arbirtray refinement level
$V_{j,k}$	kth cell of jth discretization level of Ω
I_j	index set of all cells corresponding to jth discretization level, i.e., $\Omega = \bigcup_{k \in I_j} V_{j,k}$
N_j	number of cells corresponding to discretization level j
$\mathcal{G}_j$	set of all cells corresponding to jth discretization level, i.e., $\mathcal{G}_j = \{V_{j,k}\}_{k \in I_j}$
$\mathcal{M}_{j,k}$	index set of all cells $V_{j+1,r}$ on refinement level $j+1$ resulting from the refiment of cell $V_{j,k}$, i.e., $V_{j,k} = \bigcup_{r \in \mathcal{M}_{j,k}} V_{j+1,r}$

Box Function and Box Wavelet

e	wavelet type
E	index set of all wavelet types ($e \neq 0$) and box function ($e = 0$)
E^*	index set of all wavelet types, $E^* := E \backslash \{0\}$
M	order of vanishing moments
$\tilde{\varphi}_{j,k}$	box function with respect to cell $V_{j,k}$, $\tilde{\varphi}_{j,k} \equiv \check{\psi}_{j,k,0}$ (dual scaling function)
$\check{\psi}_{j,k,e}$	box wavelet of type e with respect to cell $V_{j,k}$
$\tilde{\psi}_{j,k,e}$	box wavelet with higher vanishing moments of type e with respect to cell $V_{j,k}$
$\varphi_{j,k}$	primal scaling function
$\psi_{j,k,e}$	primal wavelet of type e with respect to cell $V_{j,k}$

Bases

$\tilde{\Phi}_j$	vector (collection) of box functions (dual scaling function), i.e., $\tilde{\Phi}_j := (\tilde{\varphi}_{j,k})_{k\in I_j}$
$\check{\Psi}_{j,e}$	vector (collection) of box wavelets of type e (dual wavelet), i.e., $\check{\Psi}_{j,e} := (\check{\psi}_{j,k,E})_{k\in I_j}$
$\tilde{\Psi}_{j,e}$	vector (collection) of modified box wavelets of type e, i.e., $\tilde{\Psi}_{j,e} := (\tilde{\psi}_{j,k,E})_{k\in I_j}$
$\check{\Psi}_j$	vector (collection) of all box wavelets, i.e., $\check{\Psi}_j := (\check{\Psi}_{j,e})_{e\in E^*}$
$\tilde{\Psi}_j$	vector (collection) of all modified box wavelets, i.e., $\tilde{\Psi}_j := (\tilde{\Psi}_{j,e})_{e\in E^*}$
Φ_j	vector (collection) of primal scaling functions, i.e., $\Phi_j := (\varphi_{j,k})_{k\in I_j}$
$\Psi_{j,e}$	vector (collection) of primal wavelets of type e, i.e., $\Psi_{j,e} := (\psi_{j,k,E})_{k\in I_j}$
Ψ_j	vector (collection) of all primal wavelets, i.e., $\Psi_j := (\Psi_{j,e})_{e\in E^*}$
$\mathbf{\Psi}^{(L)}$	primal multiscale basis, i.e., $\mathbf{\Psi}^{(L)} := \Phi_0 \cup \bigcup_{j=0}^{L} \Psi_j$
$\widetilde{\mathbf{\Psi}}^{(L)}$	dual multiscale basis, i.e., $\widetilde{\mathbf{\Psi}}^{(L)} := \tilde{\Phi}_0 \cup \bigcup_{j=0}^{L} \tilde{\Psi}_j$

Multiscale Transformation

$\hat{u}_{j,k}$	cell average of function u with respect to cell $V_{j,k}$
$d_{j,k,e}$	detail with respect to cell $V_{j,k}$ and wavelet type e
$\hat{\mathbf{u}}_j$	vector of all cell averages of level j, i.e., $\hat{\mathbf{u}}_j := (\hat{u}_{j,k})_{k\in I_j}$
$\mathbf{d}_{j,e}$	vector of all details of level j and wavelet type e, i.e., $\mathbf{d}_{j,e} := (d_{j,k,e})_{k\in I_j}$
$\mathbf{d}_j$	vector of all details of level j, i.e., $\mathbf{d}_j := (\mathbf{d}_{j,e})_{e\in E^*}$
$\mathbf{d}^{(L)}$	vector of multiscale decomposition, i.e., $\mathbf{d}^{(L)} := (\hat{\mathbf{u}}_0^T, \mathbf{d}_0^T, \ldots, \mathbf{d}_{L-1}^T)$

$m^{j,e}_{r,k}$	mask coefficient arising in two–scale relation of $\psi_{j,k,e}$
$g^{j,e}_{r,k}$	mask coefficient arising in two–scale relation of $\varphi_{j+1,k}$
$l^{j,e}_{r,k}$	mask coefficient arising in coarse–grid modification of the wavelet $\psi_{j,k,e}$
$\mathsf{M}_{j,e}$, $\mathsf{G}_{j,e}$	matrix of mask coefficients with respect to $m^{j,e}_{r,k}$ and $g^{j,e}_{r,k}$, i.e., $\mathsf{M}_{j,e} := (m^{j,e}_{r,k})_{r\in I_{j+1},k\in I_j}$ and $\mathsf{G}_{j,e} := (g^{j,e}_{k,r})_{k\in I_j,r\in I_{j+1}}$
$\mathsf{M}_{j,1}$, $\mathsf{G}_{j,1}$	composed matrix of mask coefficients with respect to all wavelets, i.e., $\mathsf{M}_{j,1} := (\mathsf{M}_{j,e})_{e\in E^*}$ and $\mathsf{G}_{j,1} := ((\mathsf{G}^T_{j,e})_{e\in E^*})^T$
M_j, G_j	composed matrix of mask coefficients with respect to all wavelets and box function, i.e., $\mathsf{M}_j := (\mathsf{M}_{j,0}, \widetilde{\mathsf{M}}_{j,1})$ and $\widetilde{\mathsf{G}}_j := (\mathsf{G}^T_{j,0}, \mathsf{G}^T_{j,1})^T$
$\mathsf{L}_{j,e}$	matrix of mask coefficients with respect to $l^{j,e}_{k,r}$, i.e., $\mathsf{L}_{j,e} := (l^{j,e}_{k,r})_{k\in I_j,r\in I_{j+1}}$
L_j	composed matrix of mask coefficients with respect to all wavelets, i.e., $\mathsf{L}_j := (\mathsf{L}_{j,e})_{e\in E^*}$
$\mathcal{M}_L$	(linear) multiscale operator
$\mathcal{M}_L^{-1}$	(linear) inverse multiscale operator

Local Multiscale Transformation

ε_j	threshold value of level j
$\boldsymbol{\varepsilon}$	sequence of threshold values, i.e., $\boldsymbol{\varepsilon} := (\varepsilon_0, \ldots, \varepsilon_L)^T$
$\mathcal{G}_{L,\boldsymbol{\varepsilon}}$	index set of local averages
$\mathcal{D}_{L,\boldsymbol{\varepsilon}}$	index set of significant details
$I_{j,\boldsymbol{\varepsilon}}$	index set of local averages corresponding to level j, i.e., $I_{j,\boldsymbol{\varepsilon}} := \{k\,;\,(j,k) \in \mathcal{G}_{L,\boldsymbol{\varepsilon}}$
$J_{j,\boldsymbol{\varepsilon}}$	index set of significant details corresponding to level j, i.e., $J_{j,\boldsymbol{\varepsilon}} := \{(k,e)\,;\,(j,k,e) \in \mathcal{D}_{L,\boldsymbol{\varepsilon}}$
$\check{\mathcal{M}}^e_{j,k}$, $\check{\mathcal{G}}^e_{j,k}$	index set of non-vanishing entries in the kth column of $\check{\mathsf{M}}_{j,e}$, $\check{\mathsf{G}}_{j,e}$ corresponding to the box wavelet
$\check{\mathcal{M}}^{*,e}_{j,k}$, $\check{\mathcal{G}}^{*,e}_{j,k}$	index set of non-vanishing entries in the kth row of $\check{\mathsf{M}}_{j,e}$, $\check{\mathsf{G}}_{j,e}$ corresponding to the box wavelet

$\mathcal{M}^e_{j,k}$, $\mathcal{G}^e_{j,k}$	index set of non-vanishing entries in the kth column of $\mathsf{M}_{j,e}$, $\mathsf{G}_{j,e}$ corresponding to the box wavelet with higher vanishing moments
$\mathcal{L}^e_{j,k}$	index set of non-vanishing entries in the kth column of $\mathsf{L}_{j,e}$ corresponding to the coarse–grid modification of the box wavelet
$\mathcal{M}^{*,e}_{j,k}$, $\mathcal{G}^{*,e}_{j,k}$	index set of non-vanishing entries in the kth row of $\mathsf{M}_{j,e}$, $\mathsf{G}_{j,e}$ corresponding to the box wavelet with higher vanishing moments
$\mathcal{L}^{*,e}_{j,k}$	index set of non-vanishing entries in the kth row of $\mathsf{L}_{j,e}$ corresponding to the coarse–grid modification of the box wavelet

Adaptive Finite Volume Scheme

$\mathbf{v}^n_{j,k}$	approximation of exact cell average $\hat{\mathbf{u}}^n_{j,k}$ at time t_n in cell $V_{j,k}$
$\mathbf{d}^n_{j,k,e}$	detail at time t_n in cell $V_{j,k}$ corresponding to wavelet type e determined by the local multiscale transformation of the local cell averages
$\lambda_{j,k}$	ratio of temporal step size τ and local spatial discretization $V_{j,k}$, i.e., $\lambda_{j,k} := \tau/\lvert V_{j,k}\rvert$
$\Gamma^j_{k,l}$	interface between cell $V_{j,k}$ and $V_{j,l}$
$\mathsf{F}^{j,n}_{k,l}$	approximation of exact flux average $\hat{\mathbf{f}}^{j,n}_{k,l}$ over the interface $\Gamma^j_{k,l}$ and time interval $[t_n, t_{n+1}]$
$\mathsf{B}^n_{j,k}$	numerical flux balance with respect to cell $V_{j,k}$
$\mathcal{D}^n_{L,\varepsilon}$	index set of significant details determined by the local multiscale decomposition of the local cell averages at time t_n
$\mathcal{G}^n_{L,\varepsilon}$	index set of local cell averages determined by $\mathcal{D}^n_{L,\varepsilon}$
$\tilde{\mathcal{D}}^{n+1}_{L,\varepsilon}$	index set of predicted significant details from the set of significant details $\mathcal{D}^n_{L,\varepsilon}$ at time t_n
$\tilde{\mathcal{G}}^{n+1}_{L,\varepsilon}$	index set of local cell averages determined by the prediction set $\tilde{\mathcal{D}}^{n+1}_{L,\varepsilon}$

Index

Editorial Policy

§1. Volumes in the following three categories will be published in LNCSE:

i) Research monographs
ii) Lecture and seminar notes
iii) Conference proceedings

Those considering a book which might be suitable for the series are strongly advised to contact the publisher or the series editors at an early stage.

§2. Categories i) and ii). These categories will be emphasized by Lecture Notes in Computational Science and Engineering. **Submissions by interdisciplinary teams of authors are encouraged.** The goal is to report new developments – quickly, informally, and in a way that will make them accessible to non-specialists. In the evaluation of submissions timeliness of the work is an important criterion. Texts should be well-rounded, well-written and reasonably self-contained. In most cases the work will contain results of others as well as those of the author(s). In each case the author(s) should provide sufficient motivation, examples, and applications. In this respect, Ph.D. theses will usually be deemed unsuitable for the Lecture Notes series. Proposals for volumes in these categories should be submitted either to one of the series editors or to Springer-Verlag, Heidelberg, and will be refereed. A provisional judgment on the acceptability of a project can be based on partial information about the work: a detailed outline describing the contents of each chapter, the estimated length, a bibliography, and one or two sample chapters – or a first draft. A final decision whether to accept will rest on an evaluation of the completed work which should include

- at least 100 pages of text;
- a table of contents;
- an informative introduction perhaps with some historical remarks which should be accessible to readers unfamiliar with the topic treated;
- a subject index.

§3. Category iii). Conference proceedings will be considered for publication provided that they are both of exceptional interest and devoted to a single topic. One (or more) expert participants will act as the scientific editor(s) of the volume. They select the papers which are suitable for inclusion and have them individually refereed as for a journal. Papers not closely related to the central topic are to be excluded. Organizers should contact Lecture Notes in Computational Science and Engineering at the planning stage.

In exceptional cases some other multi-author-volumes may be considered in this category.

§4. Format. Only works in English are considered. They should be submitted in camera-ready form according to Springer-Verlag's specifications. Electronic material can be included if appropriate. Please contact the publisher. Technical instructions and/or TEX macros are available via http://www.springer.de/author/tex/help-tex.html; the name of the macro package is "LNCSE – LaTEX2e class for Lecture Notes in Computational Science and Engineering". The macros can also be sent on request.

General Remarks

Lecture Notes are printed by photo-offset from the master-copy delivered in camera-ready form by the authors. For this purpose Springer-Verlag provides technical instructions for the preparation of manuscripts. See also *Editorial Policy.*

Careful preparation of manuscripts will help keep production time short and ensure a satisfactory appearance of the finished book.

The following terms and conditions hold:

Categories i), ii), and iii):
Authors receive 50 free copies of their book. No royalty is paid. Commitment to publish is made by letter of intent rather than by signing a formal contract. Springer-Verlag secures the copyright for each volume.

For conference proceedings, editors receive a total of 50 free copies of their volume for distribution to the contributing authors.

All categories:
Authors are entitled to purchase further copies of their book and other Springer mathematics books for their personal use, at a discount of 33,3 % directly from Springer-Verlag.

Addresses:

Timothy J. Barth
NASA Ames Research Center
NAS Division
Moffett Field, CA 94035, USA
e-mail: barth@nas.nasa.gov

Michael Griebel
Institut für Angewandte Mathematik
der Universität Bonn
Wegelerstr. 6
D-53115 Bonn, Germany
e-mail: griebel@iam.uni-bonn.de

David E. Keyes
Computer Science Department
Old Dominion University
Norfolk, VA 23529-0162, USA
e-mail: keyes@cs.odu.edu

Risto M. Nieminen
Laboratory of Physics
Helsinki University of Technology
02150 Espoo, Finland
e-mail: rni@fyslab.hut.fi

Dirk Roose
Department of Computer Science
Katholieke Universiteit Leuven
Celestijnenlaan 200A
3001 Leuven-Heverlee, Belgium
e-mail: dirk.roose@cs.kuleuven.ac.be

Tamar Schlick
Department of Chemistry
Courant Institute of Mathematical Sciences
New York University
and Howard Hughes Medical Institute
251 Mercer Street
New York, NY 10012, USA
e-mail: schlick@nyu.edu

Springer-Verlag, Mathematics Editorial IV
Tiergartenstrasse 17
D-69121 Heidelberg, Germany
Tel.: *49 (6221) 487-8185
e-mail: peters@springer.de
http://www.springer.de/math/peters.html

Lecture Notes in Computational Science and Engineering

Vol. 1 D. Funaro, *Spectral Elements for Transport-Dominated Equations.* 1997. X, 211 pp. Softcover. ISBN 3-540-62649-2

Vol. 2 H. P. Langtangen, *Computational Partial Differential Equations.* Numerical Methods and Diffpack Programming. 1999. XXIII, 682 pp. Hardcover. ISBN 3-540-65274-4

Vol. 3 W. Hackbusch, G. Wittum (eds.), *Multigrid Methods V.* Proceedings of the Fifth European Multigrid Conference held in Stuttgart, Germany, October 1-4, 1996. 1998. VIII, 334 pp. Softcover. ISBN 3-540-63133-X

Vol. 4 P. Deuflhard, J. Hermans, B. Leimkuhler, A. E. Mark, S. Reich, R. D. Skeel (eds.), *Computational Molecular Dynamics: Challenges, Methods, Ideas.* Proceedings of the 2nd International Symposium on Algorithms for Macromolecular Modelling, Berlin, May 21-24, 1997. 1998. XI, 489 pp. Softcover. ISBN 3-540-63242-5

Vol. 5 D. Kröner, M. Ohlberger, C. Rohde (eds.), *An Introduction to Recent Developments in Theory and Numerics for Conservation Laws.* Proceedings of the International School on Theory and Numerics for Conservation Laws, Freiburg / Littenweiler, October 20-24, 1997. 1998. VII, 285 pp. Softcover. ISBN 3-540-65081-4

Vol. 6 S. Turek, *Efficient Solvers for Incompressible Flow Problems.* An Algorithmic and Computational Approach. 1999. XVII, 352 pp, with CD-ROM. Hardcover. ISBN 3-540-65433-X

Vol. 7 R. von Schwerin, *Multi Body System SIMulation.* Numerical Methods, Algorithms, and Software. 1999. XX, 338 pp. Softcover. ISBN 3-540-65662-6

Vol. 8 H.-J. Bungartz, F. Durst, C. Zenger (eds.), *High Performance Scientific and Engineering Computing.* Proceedings of the International FORTWIHR Conference on HPSEC, Munich, March 16-18, 1998. 1999. X, 471 pp. Softcover. 3-540-65730-4

Vol. 9 T. J. Barth, H. Deconinck (eds.), *High-Order Methods for Computational Physics.* 1999. VII, 582 pp. Hardcover. 3-540-65893-9

Vol. 10 H. P. Langtangen, A. M. Bruaset, E. Quak (eds.), *Advances in Software Tools for Scientific Computing.* 2000. X, 357 pp. Softcover. 3-540-66557-9

Vol. 11 B. Cockburn, G. E. Karniadakis, C.-W. Shu (eds.), *Discontinuous Galerkin Methods.* Theory, Computation and Applications. 2000. XI, 470 pp. Hardcover. 3-540-66787-3

Vol. 12 U. van Rienen, *Numerical Methods in Computational Electrodynamics.* Linear Systems in Practical Applications. 2000. XIII, 375 pp. Softcover. 3-540-67629-5

Vol. 13 B. Engquist, L. Johnsson, M. Hammill, F. Short (eds.), *Simulation and Visualization on the Grid.* Parallelldatorcentrum Seventh Annual Conference, Stockholm, December 1999, Proceedings. 2000. XIII, 301 pp. Softcover. 3-540-67264-8

Vol. 14 E. Dick, K. Riemslagh, J. Vierendeels (eds.), *Multigrid Methods VI.* Proceedings of the Sixth European Multigrid Conference Held in Gent, Belgium, September 27-30, 1999. 2000. IX, 293 pp. Softcover. 3-540-67157-9

Vol. 15 A. Frommer, T. Lippert, B. Medeke, K. Schilling (eds.), *Numerical Challenges in Lattice Quantum Chromodynamics.* Joint Interdisciplinary Workshop of John von Neumann Institute for Computing, Jülich and Institute of Applied Computer Science, Wuppertal University, August 1999. 2000. VIII, 184 pp. Softcover. 3-540-67732-1

Vol. 16 J. Lang, *Adaptive Multilevel Solution of Nonlinear Parabolic PDE Systems.* Theory, Algorithm, and Applications. 2001. XII, 157 pp. Softcover. 3-540-67900-6

Vol. 17 B. I. Wohlmuth, *Discretization Methods and Iterative Solvers Based on Domain Decomposition.* 2001. X, 197 pp. Softcover. 3-540-41083-X

Vol. 18 U. van Rienen, M. Günther, D. Hecht (eds.), *Scientific Computing in Electrical Engineering.* Proceedings of the 3rd International Workshop, August 20-23, 2000, Warnemünde, Germany. 2001. XII, 428 pp. Softcover. 3-540-42173-4

Vol. 19 I. Babuška, P. G. Ciarlet, T. Miyoshi (eds.), *Mathematical Modeling and Numerical Simulation in Continuum Mechanics.* Proceedings of the International Symposium on Mathematical Modeling and Numerical Simulation in Continuum Mechanics, September 29 - October 3, 2000, Yamaguchi, Japan. 2002. VIII, 301 pp. Softcover. 3-540-42399-0

Vol. 20 T. J. Barth, T. Chan, R. Haimes (eds.), *Multiscale and Multiresolution Methods.* Theory and Applications. 2002. X, 389 pp. Softcover. 3-540-42420-2

Vol. 21 M. Breuer, F. Durst, C. Zenger (eds.), *High Performance Scientific and Engineering Computing.* Proceedings of the 3rd International FORTWIHR Conference on HPSEC, Erlangen, March 12-14, 2001. 2002. XIII, 408 pp. Softcover. 3-540-42946-8

Vol. 22 K. Urban, *Wavelets in Numerical Simulation.* Problem Adapted Construction and Applications. 2002. XV, 181 pp. Softcover. 3-540-43055-5

Vol. 23 L. F. Pavarino, A. Toselli (eds.), *Recent Developments in Domain Decomposition Methods.* 2002. XII, 243 pp. Softcover. 3-540-43413-5

Vol. 24 T. Schlick, H. H. Gan (eds.), *Computational Methods for Macromolecules: Challenges and Applications.* Proceedings of the 3rd International Workshop on Algorithms for Macromolecular Modeling, New York, October 12-14, 2000. 2002. IX, 504 pp. Softcover. 3-540-43756-8

Vol. 25 T. J. Barth, H. Deconinck (eds.), *Error Estimation and Adaptive Discretization Methods in Computational Fluid Dynamics.* 2003. VII, 344 pp. Hardcover. 3-540-43758-4

Vol. 26 M. Griebel, M. A. Schweitzer (eds.), *Meshfree Methods for Partial Differential Equations.* 2003. IX, 466 pp. Softcover. 3-540-43891-2

Vol. 27 S. Müller, *Adaptive Multiscale Schemes for Conservation Laws.* 2003. XIV, 181 pp. Softcover. 3-540-44325-8

Vol. 28 C. Carstensen, S. Funken, W. Hackbusch, R. H. W. Hoppe, P. Monk (eds.), *Computational Electromagnetics.* Proceedings of the GAMM Workshop on "Computational Electromagnetics", Kiel, Germany, January 26-28, 2001. 2003. 220 pp. Softcover. 3-540-44392-4

For further information on these books please have a look at our mathematics catalogue at the following URL: `http://www.springer.de/math/index.html`

Printing and Binding: Strauss GmbH, Mörlenbach